Automotive Cybersecurity: An Introduction to ISO/SAE 21434

Automotive Cybersecurity: An Introduction to ISO/SAE 21434

BY DR. DAVID WARD AND PAUL WOODERSON

Warrendale, Pennsylvania, USA

400 Commonwealth Drive
Warrendale, PA 15096-0001 USA
E-mail: CustomerService@sae.org
Phone: 877-606-7323 (inside USA and Canada)
724-776-4970 (outside USA)
FAX: 724-776-0790

Copyright © 2022 SAE International. All rights reserved.

No part of this publication may be reproduced, stored in a retrieval system, or transmitted, in any form or by any means, electronic, mechanical, photocopying, recording, or otherwise, without the prior written permission of SAE International. For permission and licensing requests, contact SAE Permissions, 400 Commonwealth Drive, Warrendale, PA 15096-0001 USA; e-mail: copyright@sae.org; phone: 724-772-4028.

Library of Congress Catalog Number 2021948659
http://dx.doi.org/10.4271/9781468600810

Information contained in this work has been obtained by SAE International from sources believed to be reliable. However, neither SAE International nor its authors guarantee the accuracy or completeness of any information published herein and neither SAE International nor its authors shall be responsible for any errors, omissions, or damages arising out of use of this information. This work is published with the understanding that SAE International and its authors are supplying information but are not attempting to render engineering or other professional services. If such services are required, the assistance of an appropriate professional should be sought.

ISBN-Print 978-1-4686-0080-3
ISBN-PDF 978-1-4686-0081-0
ISBN-ePub 978-1-4686-0083-4

To purchase bulk quantities, please contact: SAE Customer Service

E-mail: CustomerService@sae.org
Phone: 877-606-7323 (inside USA and Canada)
724-776-4970 (outside USA)
Fax: 724-776-0790

Visit the SAE International Bookstore at books.sae.org

Chief Growth Officer
Frank Menchaca

Publisher
Sherry Dickinson Nigam

Development Editor
Publishers Solutions, LCC
Albany, NY

Director of Content Management
Kelli Zilko

Production and Manufacturing Associate
Erin Mendicino

Contents

Preface ix
About the Authors xi

CHAPTER 1

Introduction to Automotive Cybersecurity 1

What Is Cybersecurity? 1
What Does "Cybersecurity" Mean in the Automotive Context? 3
Key Concepts and Definitions 4

CHAPTER 2

Cybersecurity for Automotive Cyber-physical Systems 7

Relationship between Cybersecurity, Functional Safety, and Other Disciplines 8
What Does "Cybersecurity" Mean in the Automotive Context? 15
The Vehicle Attack Surface 17
    Wireless Interfaces 18
        Long-Range Wireless Communications 18
        Short-Range Wireless Communications 20
    Wired Interfaces 22
    In-Vehicle Networks 24
    ECUs 25
Attack Paths and Stepping Stones 27
Addressing Cybersecurity—People, Process, and Technology 29
    Management of Cybersecurity 29
    Cybersecurity Engineering 30
    Skills Required for Cybersecurity 31
    Technology 32

CHAPTER 3
Establishing a Cybersecurity Process — 35

General Aspects of a Cybersecurity Process — 35
Standards and Best Practice — 36
Cybersecurity Lifecycle — 37
Management of Cybersecurity — 40
   Top Management Commitment — 40
   Cybersecurity Processes — 40
   Cybersecurity Culture — 40
   Roles and Responsibilities — 41
   Cybersecurity Awareness and Competence — 41
   Continuous Improvement — 42
   Information Sharing — 42

Proactive Cybersecurity Engineering — 42
   Cybersecurity Responsibilities at Project Level — 43
   Cybersecurity Planning — 44
   Concept Phase — 46
      Item Definition — 46
      Threat Analysis and Risk Assessment — 46
      Risk Treatment and Cybersecurity Goals — 47
   CAL — 48
   Cybersecurity Requirements and Controls — 49
   Design Verification — 51
   Cybersecurity Testing — 51
      Cybersecurity Testing Challenges — 51
      Cybersecurity Testing at Different Lifecycle Phases — 52
      Cybersecurity Testing Activities — 53
   Vulnerability Analysis and Management — 54

Cybersecurity during Production — 55

Reactive Cybersecurity Engineering — 56
   Cybersecurity Monitoring — 56
   Evaluation of Cybersecurity Events — 57
   Detecting and Responding to Attacks — 58
   Cybersecurity Incident Response — 58
   Assessing the Effectiveness of Detection and Response — 59
   Updates — 60

End of Cybersecurity Support	61
Decommissioning	61
The Aftermarket	61

CHAPTER 4

Assurance and Certification 63

Assurance Activities — 64
Validation — 64
Assurance Case — 65
Audit — 69
Assessment — 71
Certification — 72
Type Approval — 74

Assurance Summary — 75

CHAPTER 5

Conclusions and Going Further 77

Frequently Asked Questions — 80

What Is the Difference between UN Regulation 155 and ISO/SAE 21434?	81
To Which Types of Vehicles Does UN Regulation 155 Apply?	81
To Which Types of Organization Does ISO/SAE 21434 Apply?	81
How Do You Audit for Conformance to ISO/SAE 21434?	81
Is It Mandatory to Be Certified against ISO/SAE 21434?	82
Do I Have to Use ISO/SAE 21434 for My Cybersecurity Processes?	82
How Do I Know If My Item or Component Is Cybersecurity Relevant?	82
The Various Analysis Activities for Cybersecurity Engineering Look Very Time Consuming; How Do I Know When I Have Done Enough?	82
Does ISO/SAE 21434 Define Which Cybersecurity Tests Should Be Carried Out?	83

References — 85

Index — 91

Preface

When setting out to write this book, we specifically wanted to create a practical guide for automotive professionals who are not necessarily familiar with cybersecurity and the ISO and SAE 21434 standard to begin addressing this important new engineering discipline. Cybersecurity is, of course, much more than a set of technical engineering activities or "just" complying with a standard but affects all aspects of an organization and its products, driving the need for significant transformation activities across multiple aspects of a business. Knowing where to start and what to prioritize can be a considerable challenge in the face of an often-overwhelming level of hype and security product offerings.

We have been working in automotive cybersecurity since its beginnings as a research topic in the mid-2000s, being involved in formative activities such as collaborative projects to establish early requirements and develop cybersecurity engineering methods drawing from best practices from other disciplines and industry sectors. As the topic has grown in importance and the automotive industry has started to build cybersecurity capabilities, we have been involved from the outset in developing international best practices, standards, and regulations, including SAE J3061, ISO/SAE 21434, and UN Regulation 155. We feel this early involvement makes us well placed to provide advice not only on the final content of these documents but importantly the thinking behind the concepts and why the requirements are formulated the way they are. We bring insight into relevant practices that are not covered by the published standards and regulations, some of which were partially developed and later dropped, along with practical interpretations of the requirements.

While researching and writing the content of this book, we drew upon the various debates, controversies, and challenges that arose during the development of the standard and regulation, as well as our own experience helping clients with practical implementation of the requirements. As the development of ISO/SAE 21434 took many twists and turns, we reflected the changes as well as the background for the changes in the guidance provided in the book.

As a result, we hope the book provides meaningful guidance to get started on implementing a cybersecurity program in your organization and how to continuously improve its maturity over time.

About the Authors

Dr. David Ward is Senior Manager and leads the functional safety team at HORIBA MIRA. He is a chartered engineer with over three decades of experience in the safety, security, and reliability of automotive embedded systems.

David's professional activities include functional safety and cybersecurity training, consultancy, and assessment for automotive clients worldwide. As part leader for ISO 26262 Part 6 and a Director of MISRA, David has particular interests in software robustness and common solutions for safety and security.

David was involved in pioneering automotive industry cybersecurity initiatives such as the European EVITA project, including the development of an automotive threat analysis and risk assessment (TARA) framework. He is a nominated SAE expert to the international ISO/SAE Joint Working Group that developed ISO/SAE 21434 and is now co-leading activities to harmonize key concepts with other related standards such as ISO 26262 and the forthcoming ISO/TS 5083 on safety and security for automated driving.

Paul Wooderson is Chief Engineer and leads the cybersecurity team at HORIBA MIRA. He is a chartered engineer with over two decades of experience in embedded systems security in the automotive and smartcard domains.

Paul's professional activities include cybersecurity engineering consultancy, testing, assessment, and research and development for automotive clients globally. He is a nominated UK Expert to the international ISO/SAE Joint Working Group that developed ISO/SAE 21434, the project leader of the subsequent preliminary work item to develop the CAL and TAF concepts, and participates in the development of ISO 24089 Road vehicles—Software update engineering.

He has played a lead role in several recent cybersecurity-related collaborative research projects, ResiCAV, 5StarS, and UK CITE, and has publications on side-channel attacks and automotive cybersecurity engineering.

1

Introduction to Automotive Cybersecurity

janews/Shutterstock.com.

What Is Cybersecurity?

"Cybersecurity" is seen as a key challenge in many industries and one that consumers and regulators readily identify with due to security concerns in personal computing devices, online financial transactions, etc. In the wider contexts, terms such as "security," "IT security," and "cybersecurity" are frequently used interchangeably to refer to the need to protect and defend computer assets against malicious attacks that may lead to unwanted outcomes for the stakeholders in these assets. Typical "headline" security-related incidents that have occurred include:

- Specific deployment of malware such as the Wannacry "ransomware."
- Denial-of-service attacks such as the attack against the DNS service provider Dyn in 2016, which caused major internet websites to become unreachable.

As well as specific incidents, there is a frequent "cat and mouse" game between developers of assets and security researchers. While security researchers often follow protocols [1.1] in communicating discovered vulnerabilities to asset developers, allowing time for fixes to be deployed, the vulnerabilities are eventually published—they may then become "zero day" exploits in instances of the asset that have not been updated—meaning

©2022 SAE International 1

that malicious actors could potentially exploit them immediately on "live" assets. Examples of such vulnerabilities that have received widespread publicity include:

- The "Heartbleed" SSL vulnerability.
- The "Spectre" and "Meltdown" vulnerabilities.
- The vulnerability in WPA2 that enabled the "KRACK" Wi-Fi key reinstallation attacks.

There are many different types of malicious actors with differing motivations, and these are well documented in the literature. A common theme is that such actors actively seek out "vulnerabilities" in assets—these are weaknesses or design defects in an asset that can be exploited for nefarious purposes. Again, the literature contains detailed discussion of typical vulnerabilities, but examples include entering data into a web form that "breaks" the parsing engine allowing execution of arbitrary code or insertion of malicious data, or exploiting software defects (colloquially called "bugs") that can again be used for similar purposes.

Did Somebody Say "Cyber"?

In modern language usage, the term "cyber" has tended to be prepended to almost anything, and the realm of digital security is no different. The term originates from the ancient Greek "kubernetes," meaning the helmsman of a ship, but has been adopted over time to refer to communication and control theory ("cybernetics") by Norbert Wiener in the 1940s and then moved into popular culture with the coining of the term "cyberspace" in the 1980s.

F-Stop boy/Shuttertock.com.

The Oxford English Dictionary defines the key terms as follows:

- Cyber—Relating to or characteristic of the culture of computers, information technology, and virtual reality.
- Security—The state of being free from danger or threat.

Security is an important topic to emphasize in the current context: the discipline should be understood as "security engineering for cyber-physical systems." The subject of cyber-physical systems is continued below.

What Does "Cybersecurity" Mean in the Automotive Context?

In recent years [1.2, 1.3, 1.4], cybersecurity has become of significant interest in the automotive industry. A number of factors have combined to create this interest including:

1. The continued growth in vehicle electronics means that along with mobile devices, vehicles are now one of the major consumer-oriented software-intensive products. A modern vehicle may contain over 100 embedded microcontrollers and figures of up to 150 million lines of source code have been quoted for some vehicles. Given that typical estimates suggest that the residual defect rate in software, even when following the best development practices, is around 1–3 defects per 1000 lines of code, simple mathematics suggest software in vehicles can be an attractive proposition for security researchers to investigate.

2. Modern vehicles are connected to external computing elements. Typically, these connections take two forms—original equipment manufacturer (OEM) or consumer. OEMs integrate communications facilities for features such as remote diagnostics, remote software updates, and vehicle-to-vehicle communications. In a modern vehicle, these often take the form of a cellular modem and/or a short-range wireless communication such as Wi-Fi. Legislation may also mandate the fitment of these communication facilities for features such as E911 or eCall. Connections may also be established when consumers connect (tether) a cellular device to the vehicle. Either of these communications paths may contain vulnerabilities that an attacker could exploit to transmit malicious data to the vehicle.

3. Growth in automated driving features means that the internal communication architectures of vehicles are changing to accommodate distributed functionality. In practical terms, this means that vehicle motion functionality such as propulsion torque, braking, and steering can be commanded over the network by a number of different controllers. If an attacker can take over control of one asset in the network, then, without adequate security defenses, they may be able to request functionality from any other asset in the network.

4. Increasing levels of automation and the use of artificial intelligence mean that vehicle systems are increasingly making decisions based on data received from outside the vehicle. This data may be received via digital communications channels such as vehicle-to-vehicle communications, or via environmental sensors such as radar or cameras, which take analog measurements of the outside world. These sources are open to tampering; therefore, a means of establishing trust or confidence in the data derived from them is needed. Direct manipulation of the data itself may also be attractive to malicious actors.

Security risks are often evaluated by considering both the severity of a potential outcome (in terms of a loss that may be experienced by a stakeholder) and the likelihood that an attacker is able to exploit particular vulnerabilities in order to achieve that outcome. These concepts will be explored further in the automotive context later in this book, but it is important to note that, in general, contexts of information technology (IT) security potential losses typically relate to financial losses, loss of privacy, and operational limitations (e.g., a "denial of service" attack).

In the automotive industry, however, it should be noted that the assets are often controlling physical components in the vehicle, with the result that a security attack may also have a physical effect on the motion of the vehicle and, hence, the potential for a safety-relevant outcome, as well as the more typical outcomes associated with security incidents. Such systems are called cyber-physical systems; for this reason, it has become common practice in the automotive industry to use the term "cybersecurity" to refer specifically to the security aspects of embedded systems in vehicles that control physical components and where a security attack may therefore have a safety-relevant outcome in addition to more traditional forms of loss such as financial, privacy, and operational. More traditional approaches to security in the IT industry remain applicable in addition to the automotive context, for example, covering cellular devices, end-user application ("App") development, cloud-based services, back-office functions, and management of development data.

Information security typically centers around three security properties: confidentiality, integrity, and availability. While much focus in IT security has historically been on the confidentiality property (e.g., restricting access to data and systems), but as connected, software-driven electronic systems start to control and be influenced by physical processes, the integrity and availability properties become increasingly important.

Key Concepts and Definitions

The discussion above has already made reference to some key concepts and definitions that are used in cybersecurity, and some generic definitions are now provided. Note that specific standards and processes for cybersecurity may have their own definitions of some of these terms.

The following basic terms will now be defined: threat (scenario), asset, vulnerability, risk. These definitions are taken from the International Organization for Standardization and Society of Automotive Engineers (ISO/SAE) 21434 standard [1.5] with additional commentary.

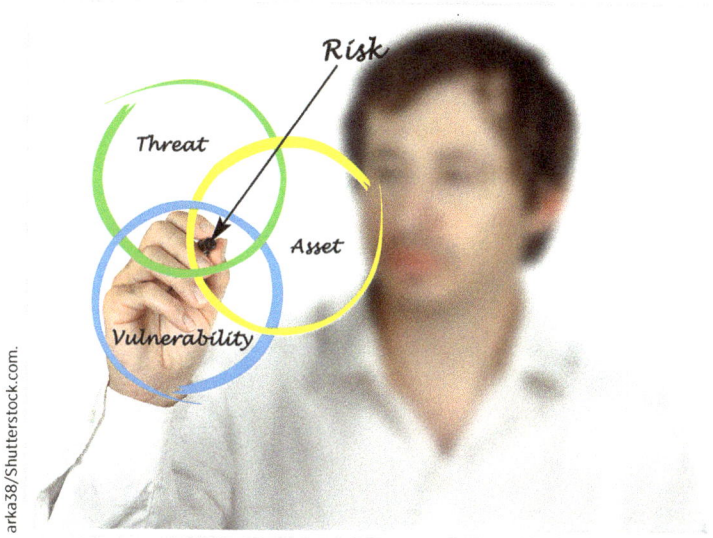

Threat scenario: "potential cause of compromise of cybersecurity properties of one or more assets in order to realize a damage scenario." This definition of threat is made specifically with respect to human-motivated actions rather than faults (although, as noted previously, malicious actors may seek to identify and exploit design faults for their purposes). Note that threat scenarios can encompass:

- Intentional exploitation of vulnerabilities with intent to cause harm.
- Intentional exploitation of vulnerabilities with no intent to cause harm but that may lead to harm (e.g., demonstrations by security researchers that unintentionally cause harm).
- Accidental compromise that is exploited by a motivated actor with intent to cause harm (e.g., personal data unintentionally leaked from a vehicle infotainment system, which is later used by criminal actors for identity fraud).

Asset: "an object that has value, or contributes to value." Note that an "asset" is viewed as having one or more cybersecurity properties whose compromise can lead to one or more damage scenarios. In the automotive industry, assets could include:

- Tangible assets such as vehicles, electronic control units (ECUs), or hardware.
- Intangible assets such as software, intellectual property, personal information (e.g., location of a vehicle or individual), goodwill, reputation.

Vulnerability: "weakness that can be exploited as part of an attack path." A vulnerability is a weakness of an asset or control that can be exploited to realize a threat scenario. Vulnerabilities can be related to technical aspects of an asset, but may also be related to process or cultural issues (e.g., social engineering of employees in order to extract sensitive information).

Risk, more specifically defined as cybersecurity risk: "effect of uncertainty on road vehicle cybersecurity expressed in terms of attack feasibility and impact." Risk is the potential that a given threat scenario will exploit vulnerabilities of an asset or group of assets and thereby cause impact to the product stakeholders. In more general risk-based approaches to engineering (e.g., in functional safety), risk is often expressed as a function of the severity of the consequence of an event and the likelihood of that consequence occurring. Terms such as "likelihood" or "probability" can be problematic to interpret in the context of cybersecurity, and this point is further considered later.

Final Thoughts

Sensevector/Shutterstock.com.

So ... what do we do about automotive cybersecurity? Chapter 2 further discusses the challenges of cybersecurity in the automotive context and the need to address cybersecurity together with related disciplines. An overview of the automotive attack surface is given and the need to address cybersecurity holistically is explained. Chapter 3 discusses the need to establish a cybersecurity engineering process based on emerging standards and best practices and how such a process can be used to build and maintain appropriate and effective security throughout the vehicle lifecycle. In Chapter 4, the legislative, regulatory, and assurance frameworks are presented. Finally, in Chapter 5 some conclusions are drawn, as well as recommendations for going further.

2

Cybersecurity for Automotive Cyber-physical Systems

metamorworks/Shutterstock.com.

In the previous chapter, we introduced the concepts of cybersecurity and of vehicles as cyber-physical systems, along with some key definitions. We now explore principles of cybersecurity for automotive cyber-physical systems in more detail.

A basic tenet in achieving cybersecurity of automotive systems is to have an approach that is founded on principles of systems engineering and risk management. Such approaches are already found in related disciplines such as functional safety, so this chapter begins with an analysis of the relationships with those disciplines. We examine the specific challenges associated with cybersecurity in the automotive domain, followed by a review of the vehicle attack surface (the points at which an attacker may seek to gain entry) and the attack paths and stepping stones (the sequence of events that an attacker may use to exploit a vulnerability).

We then describe how cybersecurity can be addressed through people, processes, and technologies.

Another important aspect is the legislative landscape. This is explored further in Chapter 4.

Relationship between Cybersecurity, Functional Safety, and Other Disciplines

A key aspect of applying cybersecurity to automotive systems is that the discipline needs to be considered in relation to other aspects of automotive engineering. Traditional vehicle engineering is based on specifying attributes or targets for the vehicle, which are usually set early on in the definition of the vehicle program, for example, by benchmarking activities against comparable vehicles. Such attributes may be objective (e.g., electric vehicle [EV] range or acceleration time to 60 mph/100 km h^{-1}) or subjective (e.g., ride and handling, NVH). These attributes are then cascaded and refined into successive levels of vehicle, system, and component requirements so that they can be implemented into the product. This "flow down" of requirements and the associated verification and validation activities are typically managed using the "V" model from systems engineering. During product development, compromises or "trade-offs" between attributes may be required (e.g., trading performance for EV range, or balancing ride comfort versus handling feel).

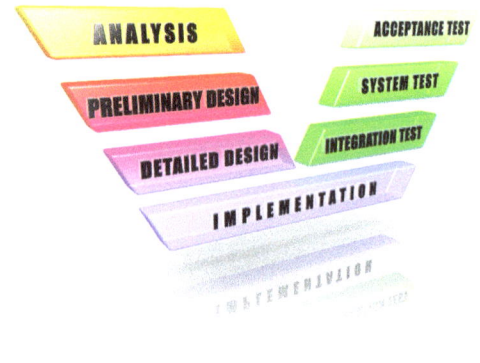

Yabresse/Shutterstock.com.

An important principle that has emerged from the discipline of functional safety is that the results of the hazard analysis and risk assessment (H&R/HARA) expressed as safety goals and their associated Automotive Safety Integrity Level (ASIL) values also effectively represent attributes that need to be defined early in the program in order to set expectations for the rigor of the product development required to achieve them.

As attributes, the safety goals and ASILs specified for a particular system ("item" in ISO 26262 terminology) therefore summarize the required performance targets and associated development rigor required and can be used by higher management to sign off on the commitment required to develop the product in accordance with the required rigor. In the authors' experience, one of the shortcomings of Edition 1 of ISO 26262 was that, due to its intentionally narrow scope, applicators tended to ignore links with other engineering disciplines and the resultant synergies that can benefit multiple domains. Example scope restrictions include:

- The definition of "hazard" is restricted to the potential harm caused by the malfunctioning behavior of electronic systems (including intersystem interactions). Other hazards such as those related to wider product safety issues and hazards inherent to the specific technology implementation are considered out of scope unless directly associated with electronic system malfunctions. For example, in the case of hazardous levels of electrical energy associated with a rechargeable energy storage system in a hybrid or EV, hazards such as electric shock and electrocution are out of scope of ISO 26262, but are in scope of other standards that set requirements for matters such as physical protection and isolation resistance. However, if an electronic system malfunction can contribute to the hazard or failure of a mitigation measure, then it is in the scope of ISO 26262.

- The definition of "harm" is restricted to injury or damage to the health of persons; wider losses or impacts are not considered even if they might be caused by malfunction of an electronic system.

This has led to a tendency to overlook or ignore significant interactions with other domains, for example, a mechanical failure mode. While the design of the product against such a failure mode is correctly addressed through standard failure mode avoidance and reliability techniques, the impact of such a failure mode on the behavior of an electronic system should be considered, for instance, to consider whether the electronic system needs to detect the failure and give a warning to the driver or implement a strategy as part of mitigation.

A case study demonstrating this point relates to the use of an electronic steering column lock (ESCL) in a vehicle. ESCL is an anti-theft provision often deployed to help meet the requirements of FMVSS 114 and the United Nations (UN) Regulation 116 for prevention of unauthorized vehicle use and replaces the traditional key-operated steering column lock in vehicles that have keyless entry and start.

From a functional safety perspective, hazard analysis would identify hazards such as undemanded locking of the steering wheel while the vehicle is in motion and failure to unlock the steering column on an authorized driver entering the vehicle, meaning that the vehicle is driven away from a parking position with the steering locked. Safety goals would be defined to avoid or mitigate the hazards and strategies derived and expressed as safety requirements to achieve these safety goals.

For the safety goal associated with avoiding undemanded locking, a typical strategy is expressed in terms of detecting malfunctions that could cause the hazard and ensuring the system is placed in a "safe state" (typically disabling of the ESCL function). For the safety goal associated with avoiding drive-away from a parking position with the steering locked, the strategy would typically also involve aspects of detecting that the steering column has not unlocked. Failure to unlock the steering column could include mechanical root causes, and while the electronic system cannot prevent these root causes, it could be required to detect that they may have occurred and to invoke an appropriate action, for example, warning the driver that the steering column is locked and that they should try to free it by turning the steering wheel and inhibiting engine start until the steering column has successfully unlocked. These latter aspects may be overlooked if a very narrow view of functional safety is taken.

Two important developments have emerged recently, and both of these have a direct bearing on automotive cybersecurity: the publication of ISO 26262 Edition 2 (ISO 26262:2018) and work on the safety of the intended functionality (SOTIF) (ISO/PAS 21448:2019). Further developments relate to legislative frameworks for cybersecurity and software updates, which are discussed in Section 2.2 on automotive challenges.

During the development of ISO 26262 Edition 2, the responsible working group was frequently asked to consider incorporating cybersecurity into that standard. It was quickly recognized that cybersecurity is a sufficient complex and unique discipline to warrant a standard in its own right; however, that, at least, some aspects of it are closely related to functional safety. Consequently, ISO 26262 Edition 2 incorporates the following four significant points related to cybersecurity (three directly and one indirectly):

1. It is now a requirement as part of establishing a safety culture in ISO 26262 Part 2 "Management of functional safety" to "institute and maintain effective communication channels" with disciplines related to functional safety of which cybersecurity is given as a specific example. This requirement acknowledges the close relationship between certain aspects of functional safety and cybersecurity. This is explored in more detail below.

2. Part 2 also contains a new informative Annex E, which provides additional guidance on interactions between functional safety and cybersecurity. This is written from the perspective of a safety practitioner to enable them to understand how decisions made from a cybersecurity perspective can impact achieving safety goals and other safety properties of an electronic system.

3. Part 6 (product development at the software level) acknowledges that, for many implementers, a common software development process is desired that collects requirements from a number of sources (e.g., safety requirements, security requirements, functional requirements) and their associated properties and attributes. Therefore, common approaches to development are desirable and may even represent common solutions (as exemplified by the recent extensions of MISRA C to cover secure coding practices in addition to safe coding practices).

4. As an indirect point, it is recognized that the classical approach of malfunctioning electronic systems being designed to "fail silent" in the cause of a malfunction is not appropriate in the case of automated driving functions at SAE J3016 Level 3 and above where the driver may not be able to react immediately (or at all), and therefore, some degree of "availability" to permit continued or alternative operation in the presence of a malfunction

may be required. Availability is indirectly linked to cybersecurity since it would need to be ensured that security solutions do not conflict with availability requirements (and acknowledging that, in so doing, they could potentially become a vector for a denial-of-service attack).

SOTIF is a new document initially published as a "publicly available specification" (ISO/PAS 21448:2019) and that will continue to be developed toward a full standard with publication in that form expected in 2022. The SOTIF approach initially arose from considering that in advanced driver assistance systems (ADAS) and automated driving systems, there are causes of hazards related to the intended function and often related to the performance of sensor systems. For instance, two of the authors' colleagues experienced "false positive" triggering of automatic emergency braking (AEB) due to a radar sensor detecting a metal plate lying on the road surface. To some practitioners, these are not considered a malfunction in the ISO 26262 sense and require a different approach. Furthermore, in considering ADAS and automated driving, there are potential scenarios and driving situations (so-called "corner" cases and "edge" cases) that are not known and which may be unsafe. Part of the approach within the SOTIF document is to consider how the extent of the domain of situations that are considered "unknown and unsafe" (otherwise known as "corner cases") can be minimized while acknowledging that this domain can never be completely understood or covered in testing. A particular significant inclusion in ISO/PAS 21448 is Table 1, an adaptation of which is reproduced below, which acknowledges that the classical approach to functional safety considering only malfunctioning behavior is only one of approximately 10 factors that can cause a system to display an unwanted and potentially unsafe behavior. Note the inclusion of cybersecurity in Table 2.1.

Considering some specific points of interaction between functional safety and cybersecurity, the following can be identified as areas of good practice:

Hazard analysis and subsequent derivation of safety requirements should consider a successful attack as a potential cause of a hazard. If threat analysis from a cybersecurity perspective identifies a potential safety-related outcome, then this needs to be communicated to the safety analysis activities and appropriately considered and aligned. This does mean an organization needs a mechanism for this communication, particularly if different teams and different processes are in use for functional safety and cybersecurity.

Furthermore, in the context of the threat analysis and risk assessment conducted in the context of ISO/SAE 21434, it is required that safety-related impacts identified during the Threat Analysis and Risk Assessment (TARA) are classified according to the ISO 26262 approach.

Related to this, as noted previously, the functional safety attributes derived from hazard analysis and risk assessment are the safety goals and their associated ASIL values. One interpretation of ASIL is that it represents a "common language" to enable communication of required rigor in the development process and the product design within the supply chain. There is general agreement that a similar approach is required for cybersecurity; at present, the concept of "cybersecurity assurance level" (CAL) is

TABLE 2.1 Safety relevant topics addressed by different ISO standards (adapted from ISO/PAS 21448:2019).

Source	Cause of hazardous event	Within scope of
System	E/E system faults (leading to malfunctions)	ISO 26262 series*
	Performance limitations or insufficient situational awareness, with or without reasonably foreseeable misuse	ISO/PAS 21448
	Reasonably foreseeable misuse, incorrect HMI (e.g., user confusion, user overload)	ISO/PAS 21448 ISO 26262 series* European statement of principles on HMI
	Hazards caused by the system technology	Product-specific standards
	System interference, e.g., connectivity issues	Interoperability (including EMC/EMI, e.g., ISO 11451 series* and ISO 11452 series*)
External factor	Successful (cybersecurity) attack exploiting vehicle security vulnerabilities	ISO/SAE 21434
	Impact from active infrastructure and/or vehicle-to-vehicle communication, external devices, and cloud services	ISO 20077 series* ISO 26262 series*
	Impact from car surroundings (other users, "passive" infrastructure, environmental conditions: weather, EMI, etc.)	ISO/PAS 21448 ISO 26262 series* Interoperability (including EMC/EMI, e.g., ISO 11451 series* and ISO 11452 series*)

* Note: "series" is used here to denote that the standard is in multiple parts.

proposed in ISO/SAE 21434, but only as informative content due to the need to gain experience in its application. There is not necessarily a mapping between ASIL and CAL (i.e., high ASIL does not imply high CAL, nor vice versa). Returning to the ESCL example, a further malfunction not mentioned previously is failure to lock the steering column when the vehicle is parked. From an ISO 26262 perspective, this malfunction would not generally be treated as hazardous since no reasonably foreseeable harm (in terms of injury to a person) can be identified, and therefore it would not have an ASIL assigned (or be treated as "QM"). However, from a cybersecurity perspective, a successful attack disabling the steering column lock could be part of the strategy of a malicious actor to steal a vehicle, and therefore it would most likely have a CAL assigned. A comparison of relative ASILs and CALs associated with the hazards of ESCL might be as shown in Table 2.2 (please note the ASIL and CAL values are indicative only).

TABLE 2.2 Comparison of relative ASILs and CALs associated with hazards of ESCL.

Hazard	ASIL	CAL (safety)	CAL (financial loss)	CAL (operational)	CAL (privacy)
Undemanded locking	High	High	High	High	None
Failure to lock	None	None	High	None	None
Failure to unlock	Low	Low	High	High	None
Undemanded unlocking	None	Low	High	None	None

From Table 2.2, it should be observed that:

- An attribute such as CAL may incorporate different scorings for different adverse security outcomes (e.g., security, financial loss, operational limitation, or loss of privacy), and as such, it can be considered as a vector quantity rather than a single value.

- For the example of this particular system, there is a strong correlation between ASIL and the safety aspect of CAL, although this may not be the case for all systems.

- The financial aspect of CAL is "high" for failure to lock since this may be associated with theft of a vehicle, even though this failure is not considered safety-relevant from an ISO 26262 perspective.

- The financial and operational aspects of CAL are "high" for failure to unlock. Failure to unlock may expose the vehicle owner to costs of alternative transport and repairs. It also represents an operational limitation that, while frustrating for the owner, may have wider implications for the manufacturer if it is perceived that a specific product or range of products are vulnerable to such an attack.

- This last observation also raises the question of whether the impact is addressed from the perspective of the vehicle user and an individual instance of the product, or whether implications for the fleet of a product and/or the manufacturer need to be considered. ISO/SAE 21434 has been written from the former perspective, but users of the standard are able to expand the scope of impact should they desire.

The alignment and possible conflict between safety requirements and security requirements has to be considered, for example, it must be ensured that a safety concept is not a source of a potential security vulnerability. An example could be a safety mechanism that removes a function (potentially leading to reduced performance or even stopping of a vehicle) that could be exploited by an attacker whose goal is to stop the vehicle in order to threaten the occupants or steal the vehicle. Conversely, a security concept should not be a cause of not achieving a safety requirement (e.g., a requirement for encryption and its associated processing time overhead may conflict with a time-sensitive fault detection mechanism from a safety perspective; a real-time monitoring and incident response mechanism may conflict with a safety-related availability requirement).

In specification, design, and implementation of the system architecture from a functional safety perspective, the properties of "independence" and "freedom from interference" have to be specified and achieved. "Independence" between architectural elements is generally required where it is desired or necessary to have a redundant implementation, and ASIL decomposition can be applied as a result. "Freedom from interference" is required where safety-related and non-safety-related elements have to coexist in the same architectural context, and it has to be demonstrated that a failure in the non-safety-related element cannot have an adverse impact on the safety-related element. Similar concepts will be required from a security perspective, such as demonstrating "isolation" (or a similar property) between security-related and non-security-related components. As a further factor, the overlap and interaction with safety architectural constraints may be required. The diagram below shows an example of a high-level software architecture comprised of components (sometimes known as partitions) and where required properties such as freedom from interference, independence, and isolation may all be required within the same software architecture. Note that not all the required properties are shown in this Figure 2.1.

Note that in the context of cybersecurity, isolation is required between security-relevant and non-security-relevant components (e.g., between software component Y and software component U). From a functional safety perspective, it has to be shown that there is not a cascading failure from software component Y that could affect a safety requirement allocated to software component U; however, from a security perspective, it also has to be shown that software component U cannot have an adverse security impact on software component Y. Isolation is also required between security-relevant components that carry different assurance levels, e.g., between software component Y and software component W.

Generally speaking, the safety mechanisms and countermeasures that enforce the required properties will have to be developed to the highest integrity and assurance requirements of the components that they protect; in the example above, an

FIGURE 2.1 Example of freedom from interference and independence in a software architecture.

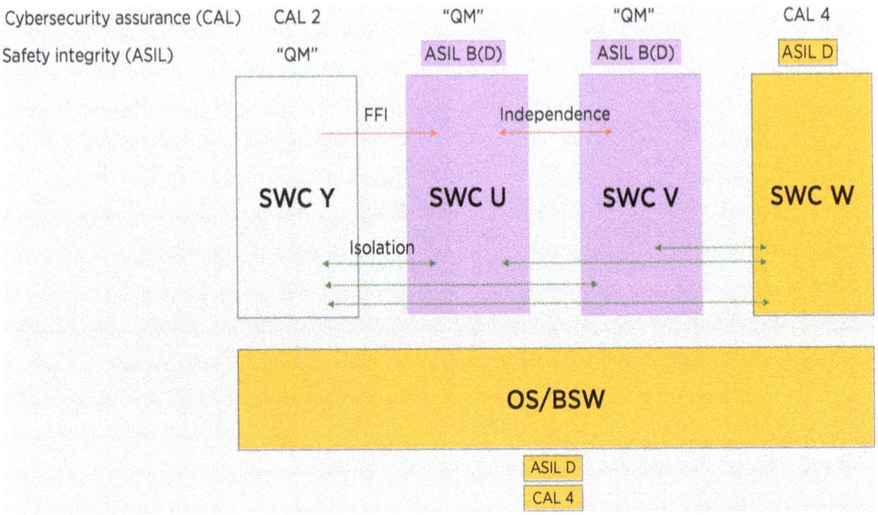

operating system (OS) or basic software layer, therefore, inherits safety integrity requirements of ASIL D and cybersecurity assurance requirements of CAL 4. While some of the properties may align (e.g., software component W that has both higher safety integrity requirements and cybersecurity assurance requirements compared to software component Y, and there may be common solutions such as memory protection), this is not always the case.

Although the above topics relate specifically to the interactions between functional safety and cybersecurity, the achievement of cybersecurity may also require interactions with and trade-offs with other attributes such as quality, reliability, and maintainability. Ultimately, approaches are required based on systems engineering and using an industry-agreed method of risk prioritization such as CAL.

What Does "Cybersecurity" Mean in the Automotive Context?

From the foregoing discussion on the interactions between cybersecurity and other disciplines, it is evident that there are some specific challenges associated with achieving cybersecurity in automotive products. Some specific challenges include:

- Safety-related implications.
- Solutions needed that scale to vehicle platforms and respect other attributes (such as availability).
- Vehicle lifetimes.
- Changing models of vehicle ownership.
- How and when to update software in vehicles.
- "Right to repair" legislation.
- End of life.
- The "software-defined car."
- Legislation.

These are discussed further below with the exception of safety-related implications, which are already covered above.

Scalability—security solutions need to scale to vehicle applications respecting domain needs such as limited resources (memory, processing capacity) and the real-time nature of some of the applications. This is closely related to the need to consider wider attributes of functions such as availability or a safety-related timing property.

Vehicle lifetimes—the typical design life of a passenger car is approximately 10 years (as a rough order of magnitude). While there are moves around more frequent software updates (q.v.), the hardware is effectively fixed for the lifetime of the vehicle. As a consequence, manufacturers are proposing more flexible electrical architectures with general-purpose hardware capable of supporting a wider range of software-defined features, which could therefore be updated during the lifetime of the vehicle.

Changing models of vehicle ownership—in many established markets for road vehicles, most vehicles are owned (whether as an outright capital purchase or through a finance arrangement) by an individual and used solely by them. As a consequence, many vehicles spend a significant proportion of their life inactive and may only be used for 1 to 2 hours per day. When the vehicle is in use, the owner is not particularly concerned about sharing personal data. In contrast, new models of shared ownership have been proposed where users would rent a vehicle on a "power by the hour" scheme, and consequently, the need to protect the personal data of different users will become important. A micro-example of this is seen in rental cars, wherein the authors' experience the in-car entertainment system frequently contains an extensive history of past paired devices, which potentially could lead to leakage of data between users.

How and when to update software in vehicles—while mechanisms for in-service software updates in vehicles have existed for many years, these are frequently performed by a dealer during a maintenance action. However, these do not guarantee a high level of coverage of the fleet of a particular vehicle as they are dependent on users presenting a vehicle for maintenance. Even a "mandated" action, such as recall, is not 100% effective. Conversely, many readers will be familiar with frequent software updates in desktop computers and similar devices, which often have security-relevant changes. Some vehicle manufacturers have already started to use "over the air" (OTA) software updates, often to deploy new features to the vehicle, but which are also attractive to deliver security-related updates in a more rapid timeframe compared to traditional update mechanisms.

Software updates bring a number of challenges directly associated with cybersecurity and the wider issues mentioned above. While software updates provide an efficient and effective mechanism for performing security updates, it also needs to ensure that only authorized personnel can make the updates; otherwise, the update mechanism itself could become a vulnerability. There are potential availability considerations, for example, certain safety-related functionality should not be updated while the vehicle is in use; however, if such updates require the vehicle to be stationary, can it be guaranteed they will be completed before the user requires the vehicle again?

The "software-defined car" is a term that has recently started to be used. As noted above, some manufacturers are starting to deploy architectures with a greater emphasis on a flexible hardware platform that can support a variety of different applications through the deployment of software and even "App Store" models where additional functionality can be provided in the aftermarket either from the OEM, its supply chain, or third-party providers.

Legislation—recently work has taken place in the United Nations Economic Commission for Europe (UNECE) world forum for harmonization of vehicle regulations to prepare regulations for cybersecurity and software updates. These regulations entered into force in January 2021. The motivations for the cybersecurity regulation should be self-evident, but demonstrate that regulators are taking a specific interest in this aspect of vehicles (whereas a more "hands-off" regulatory approach to functional safety has been evident up to now). The motivations for the software update regulation are related to the points discussed above (the desire for more rapid updates, but balancing this against safety and security considerations). A significant aspect of both of these regulations is that they require a third-party audit of the management system used by the manufacturer to address engineering rigor in product development from the cybersecurity and software update perspectives. This topic is further discussed in Section 4.1.6.

Another interesting regulatory aspect concerns consumer choice legislation such as "right to repair" in the USA and elsewhere. Essentially, such legislation is aimed at allowing vehicle owners the choice of repairers and not to be required to use a manufacturer's own service network. In turn, this requires that manufacturers provide access to information and tools to third-party repairers on the same basis as their own agents. The equipment, tools, and information that is made available in this way could have the potential for illegitimate use in the wrong hands, for example, service tools that permit reprogramming of access and usage controls, such as a keyless entry system, still need to be subject to security controls.

A related aspect is the Digital Millennium Copyright Act (DMCA) and the right to "jailbreak" products and conduct security research on them. In the 2015 revision of this Act, legislators were successfully lobbied to grant exceptions related to software in road vehicles.

One specific exemption reads "Computer programs that are contained in and control the functioning of a motorized land vehicle such as a personal automobile, commercial motor vehicle or mechanized agricultural vehicle, except for computer programs primarily designed for the control of telematics or entertainment systems for such vehicle, when circumvention is a necessary step undertaken by the authorized owner of the vehicle to allow the diagnosis, repair or lawful modification of a vehicle function; and where such circumvention does not constitute a violation of applicable law…"

This is an interesting case study, as the exemption was granted to permit owners to conduct maintenance or modification on vehicles (provided this did not violate laws, such as emissions legislation). However, such an action may have an unforeseen outcome, but this is the result of an accidental action rather than a malicious action. Often "cybersecurity" issues are seen as related to deliberate, malicious action, but it is also necessary to consider unwanted and unforeseen outcomes of this type of use case.

The DMCA also contains an exemption related to "good faith" security research, which again has the potential for an unforeseen outcome as the result of an accidental action.

End of life—when a vehicle reaches the end of its life (including change of owner), it is important to ensure that personal data is erased. Use cases where personal data needs to be erased include replacement of a component, change of ownership, shared-use vehicles (including rental cars), and at end of life of a vehicle where it is decommissioned, dismantled, and disposed.

The foregoing discussion has demonstrated some of the specific challenges that need to be considered in addressing and achieving automotive cybersecurity. We return later to explain how these can be addressed through processes (in Chapter 3) and assurance activities (in Chapter 4).

The Vehicle Attack Surface

Increasing electronic content, connectivity, and automation of vehicles means that there is a large and diverse attack surface. The attack surface can be described as the set of entry points through which attackers may initiate attacks on the vehicle to eventually compromise assets within the vehicle systems.

The attack surface can be categorized based on the type of access required of the attacker:

- Wireless interfaces.
- Wired interfaces.
- In-vehicle networks.
- ECUs.

Wireless Interfaces

Today's multitude of vehicle connectivity features expose a number of potential entry points for attackers. These features provide the vehicle with communication capabilities over both long and short ranges. It is important to remember that the specified range of wireless technology is only the minimal range that is specified to be supported by that technology. Attackers are not bound by specified operating ranges and can potentially extend the nominal range of a wireless communications system by the use of adapted equipment such as a specifically tuned antenna or amplifiers.

ADAS and automated driving functionality also require the use of sensing technologies, which also constitute a wireless attack vector. The threats posed by these types of wireless interface are described in the remainder of this section.

Long-Range Wireless Communications

jamesteohart./Shutterstock.com.

Connected car services mean that many vehicles are now equipped with on-board cellular modems, which connect via mobile networks and the internet to the vehicle manufacturer's backend systems. This provides the means for attackers to potentially access one or multiple vehicles from a remote location, possibly on the other side of the world. Attacks have been demonstrated [2.1] that use a fake cellular base station to convince a vehicle to connect to a fake backend server, for example, a server masquerading as a vehicle manufacturer's OTA software update server. The cost of implementing such an attack is now within the reach of even attackers with limited resources as a fake base station can be implemented using a low-cost software-defined radio (SDR) with freely available software [2.2].

"Vehicle to everything" (V2X) communications also provide a vehicle with a shorter range communication channel to other vehicles and roadside infrastructure with transmission distances of hundreds of meters. V2X communications will become more prevalent in the coming years and consist of two technologies, which are likely to coexist: The first of these is dedicated short-range communications, a short-range radio specification based on IEEE 802.11p, which is a derivative of the Wi-Fi specification optimized for the vehicle environment. The second is cellular V2X or C-V2X, based on LTE and currently under development by 3GPP [2.3].

The security of V2X communications has been an active research topic for a number of years. Threats against V2X communications include the injection and tampering of messages on the air interface. Although the first use cases of V2X will be information services, in the future the technology is expected to be used for more critical purposes, including safety-related functionality. In this case, spoofed or tampered V2X messages could pose a risk of harm to vehicle occupants and other road users. Another V2X threat is the eavesdropping of messages for the purpose of tracking vehicles, compromising the privacy of occupants. With these threats in mind, V2X security standards are already available from both Institute of Electrical and Electronic Engineers (IEEE) and European Telecommunications Standards Institute (ETSI) [2.4]. These standards define security requirements for V2X communications, including a secure messaging format based on cryptographic digital signatures and a privacy-protecting public key infrastructure to manage the keys. Field trials are currently underway in a number of countries to test the security specifications, as well as the technology itself.

A further wireless interface is the Global Navigation Satellite System (GNSS) receiver, which is used to receive satellite signals, which are used to obtain position and time information. GNSS variants are susceptible to a number of threats, including jamming (blocking the signal by transmitting a stronger signal, which prevents the genuine signal from reaching the receiver). GNSS signals can also be spoofed by creating fake signals at GNSS frequencies, which mimic the constellation of satellites and cause the receiver to derive plausible but incorrect position or time information. Proposed intelligent transport and automated driving functions rely on accurate and reliable location data, but the inherently low-power GNSS signals are susceptible to interference from simple low-cost jamming and spoofing systems.

A GNSS spoofing attack against an automated driving system, which induces a small error in the position interpreted by the GNSS receiver, could potentially lead to the system making an incorrect driving decision. The exploitability of such an attack depends on the measures taken in the automated driving system, for example, the fusion of measurements from different sensors, plausibility checking, and how much reliance is placed on the use of GNSS.

Proposed V2X communications specifications [2.5] include security requirements for the generation and verification of digital signatures to ensure the authenticity of messages sent between vehicles and the roadside infrastructure. These mechanisms include checks on the validity periods of the digital certificates in order to prevent misbehaving nodes to exist in the system for long periods. If a communications unit relies on GNSS as a source of date/time, an attacker could use GNSS spoofing attacks to modify the system time and subvert such a certificate validity check, further enabling the spoofing of V2X messages.

The feasibility of both jamming and spoofing attacks has increased in recent years with the availability of low-cost SDR and freely available software and configuration data to create the jamming or spoofing signals.

A final long-range wireless technology to consider is broadcast radio. Attacks have been demonstrated against both Frequency Modulation (FM) and Digital Audio Broadcasting (DAB) radio [2.6], which range from relatively harmless manipulation of the radio display to code injection via a vulnerable DAB software stack. The broadcast nature of FM and DAB radio makes attacks on them highly scalable, with the possibility that a single attack can affect many vehicles within range of the attacker's broadcast.

Short-Range Wireless Communications

Alexander Supertramp/Shutterstock.com.

A variety of shorter-range wireless technologies in vehicles also pose threats. Most mainstream vehicles are now fitted with Bluetooth connectivity for pairing occupants' mobile devices wirelessly to the vehicle's infotainment system. This enables functionality such as music playback and telephony, but as with all wireless interfaces, this represents a potential entry point for attackers. The Bluetooth communications stack is relatively complex and, therefore, a likely source of implementation vulnerabilities that could be exploited by an attacker to gain control over the host system to mount further attacks. In addition, some Bluetooth features may offer privileged access to the system, such as the ability to mount the filesystem remotely [2.7].

Wi-Fi technology is also increasingly found in vehicle systems, offering either wireless hotspot functions for the vehicle occupants or Wi-Fi clients to allow the vehicle to connect to other Wi-Fi access points, for example, to enable the download of software updates for systems within the vehicle. If not properly secured, these interfaces could provide a means for attackers to download malicious software updates to the vehicle or to gain remote access to other systems in the vehicle [2.8]. Examples of attacks on Wi-Fi functionality include circumvention of the Wi-Fi security mechanisms such as WPA 2 PSK [2.9] and eavesdropping Wi-Fi communications to fingerprint or track vehicles and their occupants, potentially tracking them over time.

A number of wireless systems operating in the unlicensed frequency spectrum also present potential attack vectors. For example, tire pressure monitoring systems (TPMS) typically provide a simple radio link operating on 433 MHz or 315 MHz to transmit tire pressure measurements from the tires to a central ECU in the vehicle. It has been demonstrated in [2.10] that this wireless link can be eavesdropped and the IDs within the transmitted data packet used to uniquely identify and track vehicles, leading to a privacy risk for the occupants. The same authors also demonstrated packet spoofing attacks, enabling false low-pressure warnings to be displayed to the driver. Although TPMS are not usually safety-related systems, this attack vector could potentially enable an attacker to convince the driver of a target vehicle to stop at the roadside with possible indirect consequences for the safety of the vehicle occupants.

Remote keyless entry systems operate on similar frequencies, and the means of vehicle theft have continuously been adapted to defeat increasingly sophisticated security measures incorporated in such systems. Early remote keyless entry systems relied on a single, constant code transmitted from the keyfob to the vehicle. These could be relatively easily subverted in a replay attack in which a car thief eavesdrops on the wireless channel, "grabs" the code, and replays it later to unlock the vehicle. Security measures introduced to prevent such replay attacks have been implemented for many years, such as rolling codes and encryption; however, a number of researchers have demonstrated methods to overcome these measures [2.11].

The move toward passive keyless entry systems, in which the vehicle user does not have to press a button on the keyfob but can simply walk up to the vehicle and unlock it, has introduced the possibility of relay attacks [2.12]. These attacks involve attackers working in pairs, each with a radio relay device that can convert the low-frequency signal from the vehicle to the key to a higher frequency signal that can be transmitted over a longer distance. The attack method takes advantage of the invalid assumption that the ability for the keyfob to communicate with the vehicle implies that the keyfob is close to the vehicle.

ADAS and automated driving systems make use of sensors on the vehicle to make measurements of the surrounding physical environment, which are then used to make driving decisions, such as steering or braking. Examples of such systems include Lane Keeping Assistance and AEB.

Although such sensors are making physical measurements rather than transmitting or receiving data packets, they must still be considered as a potential vector through which an attacker can manipulate the behavior of the vehicle. Sensor technologies such as ultrasonics, radar, and lidar are vulnerable to both jamming and spoofing attacks, which can be implemented remotely and could result in manipulation of the behavior of automated vehicles. Attacks have been demonstrated that spoof the measurements made by ultrasonic and radar sensors [2.13, 2.14], enabling an attacker to falsely convince the vehicle that a fake object is present or that a real object is not present. Research has also shown that it is possible to blind cameras using a simple laser pointer and spoof lidar signals to make objects appear closer or further away from the vehicle than they really are [2.15].

Wired Interfaces

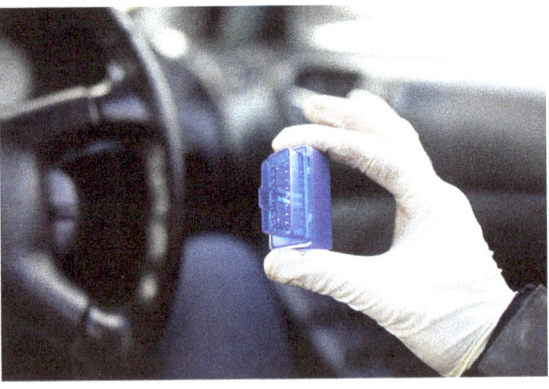

The physical interfaces of a vehicle also present potential entry points for an attacker. Although attacks mounted via these interfaces place greater demands on the attacker in terms of the required level of access to the target vehicle, such attacks are relevant in a number of use cases such as car sharing, vehicle rental, valet parking or service, and maintenance scenarios. In these scenarios, the attacker may have temporary access to either the exterior or interior of the vehicle for sufficient time to carry out an attack or to tamper with one or more vehicle systems such that an attack is deployed at a later time.

Physical interfaces accessible from the exterior of the vehicle include the electric charging connector used to connect an EV to a charging point. This interface can potentially be used to mount attacks directed at the vehicle or from the vehicle to the charging infrastructure or grid. Such attacks may have impacts on safety and availability but also potential financial implications since EV charging typically involves payment transactions. Research has demonstrated ways in which the EV charging ecosystem might be attacked [2.16], and while approaches to address security in EV charging-related international standards [2.17] exist, further industry efforts are underway to further investigate cybersecurity implications and develop more comprehensive mitigation approaches [2.18].

Inside the vehicle, many other physical interfaces exist, some of which are standard communication interfaces also found in consumer electronics such as Universal Serial Bus (USB) or optical media drives connected to infotainment systems. Researchers have demonstrated [2.19] that it is possible to use a malicious music file on a compact disk (CD), which, when played in the vehicle's CD player, exploits software vulnerabilities in a vehicle infotainment system to cause unintended effects on the vehicle. The widespread use of USB to provide functionality such as updates to navigation maps and system software exposes vehicle systems to similar attacks to those of other IT systems with USB interfaces, such as the "BadUSB" attacks [2.20].

Diagnostic connectors such as the On-Board Diagnostics (OBD)-II port [2.21] also provide a potential entry point for attackers. The OBD-II port is typically used during maintenance to perform diagnostic functions on the vehicle electronic systems, which could be manipulated by an attacker to gain unauthorized access to particular systems or data or to make unauthorized modifications to the data stored in a system. Diagnostic access from the OBD-II port to ECUs is typically via the Controller Area Network (CAN) bus and increasingly via other network technologies. Without sufficient measures to restrict the processing of network messages originating from this port, an attacker may be able to inject fake network messages that control vehicle systems such as steering or brakes [2.22, 2.23].

Researchers have demonstrated attacks that exploit weaknesses in diagnostic functions implemented using Unified Diagnostic Services [2.24] or Diagnostics Over Internet Protocol. For example, in [2.8] the authors describe how vehicle ECUs can be put into a diagnostic state while in motion, causing loss of steering assistance or braking. Other vulnerabilities include insufficient authentication of certain invasive diagnostic functionality, such as writing ECU memory or downloading software. Commonly used authentication mechanisms, such as seed-key algorithms based on challenge-response method, have been found to be implemented using easily predictable keys and seed values with insufficient entropy.

Increasingly, the OBD-II port is being used to connect aftermarket devices to the vehicle with in-built wireless connectivity, such as "pay as you drive" or "black box" insurance dongles and telematics units, which connect to third-party services via a cellular link. Several vehicle manufacturers also offer aftermarket connected car services enabled by fitting a similar type of dongle to the OBD-II port. The wireless interface provided by such dongles presents another potential entry point for attackers to mount remote attacks on the vehicle, which may have safety implications if the wireless interface enables an attacker to interact with other on-board ECUs via the vehicle network.

In-Vehicle Networks

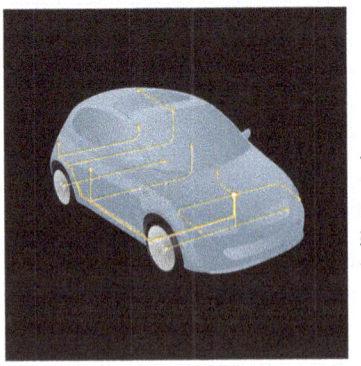

In modern vehicles, data and control signals are passed between the electronic systems by means of in-vehicle networks, which reduce the quantity of wiring that would otherwise be required to connect the relevant systems directly. At the current time, several in-vehicle network technologies are in use including CAN, Local Interconnect Network (LIN), FlexRay, and Automotive Ethernet.

CAN is currently the dominant network technology in most types of vehicles and has been widely studied in terms of its susceptibility to message injection attacks. The CAN specification was developed in the 1980s before vehicles were equipped with external connectivity, and as such, the in-vehicle network could be considered a closed system. It was therefore designed as a differential two-wire bus optimized for reliability and robustness to interference; intentional manipulation or insertion of messages was not part of the design criteria. As a result, CAN messages do not contain identifiers for the sender or receiver, and therefore, no verification of the authenticity of the message is possible. Furthermore, integrity protection is specified in the form of a CRC, which is effective at detecting accidental corruption of messages in transit, but entirely unsuitable for detecting deliberate manipulation. Researchers have demonstrated how these properties can be exploited to inject messages onto the CAN bus masquerading as a genuine ECU [2.22]. This can be achieved either via connection to and tampering of the CAN bus itself, via a diagnostic connector such as OBD-II or direct connection to a CAN wiring harness, or by exploiting vulnerabilities in an ECU connected to CAN, such that it can be taken over and forced to send malicious messages.

Attacks on other vehicle buses such as LIN and FlexRay have been studied [2.25], but are less widely documented compared to CAN. LIN is typically used for connecting body functions such as window switches or lighting to the relevant body control ECUs, and as such, the impacts of tampering with LIN messages are generally lower. FlexRay is often used for communication between safety-related systems due to its time determinism and robustness; it is, therefore, necessary to ensure that FlexRay messages cannot be tampered with, spoofed, or blocked, any of which could be a means for an attacker to cause unintended operation of a safety-related system such as braking or steering. The FlexRay protocol is more complex than CAN, with time synchronization being a major requirement; this makes the injection of messages more difficult as the attacker must ensure that any injected messages are inserted in the correct time slot.

Automotive Ethernet is emerging as a key in-vehicle network technology with multiple uses in the vehicle due to its bandwidth and flexibility. Specific automotive variants and extensions to the standard IEEE Ethernet protocols exist or are under development [2.26], including time-triggered protocols, which enable applications requiring time determinism to make use of Ethernet. Since Automotive Ethernet is based on standardized protocols widely used in IT networking, the technology is potentially susceptible to similar attacks as IT networks, with the associated wide availability of information and tools to enable such attacks. It also means that many of the countermeasures that can be applied to Ethernet-based IT networks are also suitable for Automotive Ethernet applications, with some appropriate modification.

Often there will be multiple technologies in use in a single vehicle, for example, CAN for safety-related systems, LIN for body control functions, and Automotive Ethernet for backbone connections between domain controllers.

ECUs

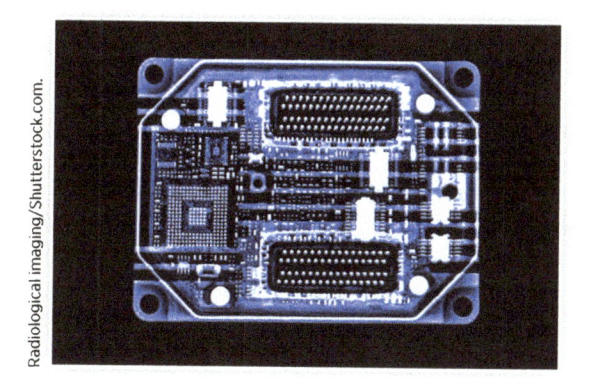

Radiological imaging/Shutterstock.com.

As well as the attack vectors presented by vehicle-level interfaces and networks, it is also important to consider the systems within the vehicle, comprising the ECUs, sensors, and actuators. The functionality of the various vehicle systems is implemented in these elements, which if compromised can lead to failure of those functions or unintended side effects of their operation.

The technology on which the ECUs within a typical vehicle are based is highly heterogeneous, with ECUs ranging from relatively simple microcontroller-based systems with minimal interfaces and "bare metal" software performing a dedicated function to complex system-on-chip (SoC)-based systems with one or more rich OS such as Linux. Such advanced systems may even run hypervisor or separation kernel software with several different OS on the same unit, for example, a Linux OS and AUTOSAR-based OS.

This heterogeneity increases the complexity of the attack surface presented by the ECUs of a vehicle in that each may present a different set of attack vectors, unlike a network of personal computers all based on compatible hardware, common OS, and application software.

For example, an infotainment system typically consolidates entertainment and informational functions, such as radio, media playback, and navigation, and as such presents a wide range of interfaces. These interfaces include connections to other vehicle systems via in-vehicle networks like CAN and Ethernet, as well as interfaces intended to connect to external systems or devices, such as USB, Bluetooth, and Wi-Fi. Media such as CD or DVD can also be a means through which attacks are delivered, as described in section 2.3.2. To provide the increasingly demanded range of audio, video, and communications functions, infotainment systems typically embed at least one complex SoC (sometimes multiple SoCs) running a rich OS such as Linux, as well as one or more microcontrollers to control interfaces with the in-vehicle network. This introduces a large software and hardware attack surface with several potential hardware attack vectors such as exposed debug interfaces and serial ports. These ports are typically implemented for the purposes of testing during development, production, or field return. However, if unprotected they can enable an attacker to extract or modify the ECU embedded software, for example, reading and writing the software via JTAG.

Conversely, ECUs with dedicated vehicle control functionality, such as antilock braking systems or electric power steering systems, are typically based on much simpler system architectures, consisting of single microcontrollers with a lighter OS (e.g., AUTOSAR based) or even "bare metal" software.

The ECUs are also where many cybersecurity controls are implemented to prevent, detect, or respond to attacks on the ECUs themselves or on communications over in-vehicle or external networks. These controls include cryptographic functionality for the encryption of data and authentication of messages or entities, as well as redundancy mechanisms and plausibility checks, to provide robustness against corrupted communications or sensor data. Attackers may attempt to subvert these controls to render them ineffective in order to mount the original attack the controls were intended to protect against, for example, an attacker attempting to inject fake messages on a network protected by an authentication mechanism, such as a digital signature, may first try to attack the authentication mechanism to either bypass it or deduce a secret authentication key, which can then be used to forge messages.

Attacks to subvert such mechanisms include:

- *Side-channel attacks*, in which the attacker exploits leakage of data processed by an electronic system through physical characteristics of that system, such as transient power consumption or electromagnetic emissions, rather than through its functional interfaces. These attacks are often used to disclose sensitive data such as cryptographic keys during the execution of a cryptographic algorithm and make use of the fact that the leaked data is correlated with the data processed by the device. The leakage through these unintentional channels is measured, digitized, and then analyzed using signal processing and statistical techniques. Side-channel attacks have been published against many cryptographic algorithms, including AES, RSA, elliptic curve-based algorithms, and certain types of keyed hash functions. Many variants and enhancements of these side-channel analysis techniques have been published in the literature and continue to be developed by an active research community.

- *Fault injection attacks* manipulate the behavior of semiconductor hardware by deliberately injecting faults. Faults may be injected by a number of physical means, including, but not limited to,
    - Operating the system outside of its specified operating range, for example, setting the supply voltage too high or too low or heating or cooling the device.
    - Applying short duration glitches to power supply or clock pins.
    - Generating electromagnetic pulses in close proximity to the device.

Such faults would usually lead to a failure of some functionality of the hardware, for example, skipping or modifying an instruction that is executed by a microcontroller. This failed operation can then be exploited by the attacker to bypass a cybersecurity control, for example, PIN authentication function.

Attack Paths and Stepping Stones

Attackers may be able to realize their objectives by deploying a single type of attack in isolation or by combining several types of attack into an attack path against a particular system.

For example, an attacker may try jamming a wireless communications channel by transmitting an arbitrary signal at the relevant frequency and bandwidth and sufficiently high power to prevent a wireless communications signal from reaching a vehicle, which may be enough to cause damage or harm.

Alternatively, as demonstrated by Miller and Valasek [2.8] in 2015, it may be necessary to identify several vulnerabilities in both the vehicle systems and in the wider connected infrastructure and then attempt to chain them together in order to realize a complete attack path and cause damage or harm. In this example, the authors were able to devise a remotely executed attack, exploiting a set of vulnerabilities that could cause the vehicle to make an unintended maneuver with the potential to cause physical harm. Furthermore, these research findings indicated how, having developed the attack on a single vehicle, it could be scaled to attack multiple vehicles of the same specification.

FIGURE 2.2 Attacker economics.

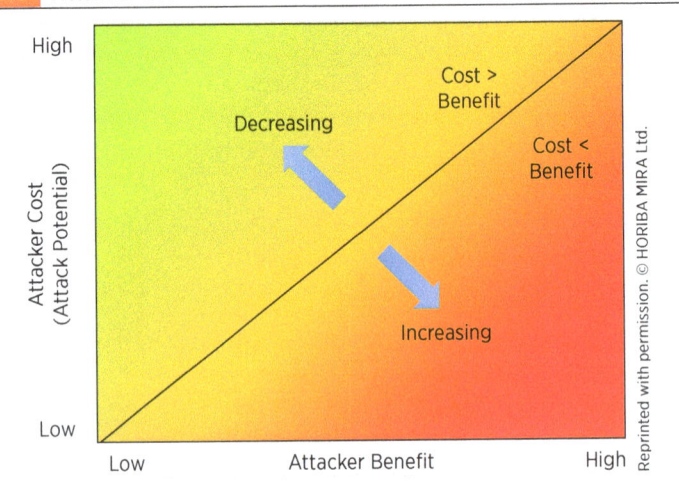

Attack paths can vary in complexity, cost, and difficulty to execute, and attackers will always look for the easiest and lowest-cost way to achieve their objectives, considering the cost-benefit balance of attacking a system. The likelihood of a particular attack being attempted is influenced at least as much by the attacker's motivation as by the difficulty/feasibility of an attack: an attacker is unlikely to attempt even a highly feasible attack if there is no significant "payback," i.e., motivation, as illustrated in Figure 2.2. Conversely, an attacker who sees a very high payback will be prepared to incur the cost of a very difficult, low feasibility attack if necessary. This is why industry sectors like payment cards and pay TV place significant effort on mitigating against high-effort attacks.

As vehicles become increasingly automated and connected to a wider array of other systems within the mobility ecosystem, the potential attack surface increases, as do the potential impacts of a successful attack. Intentional interference with connected and automated vehicles, and the wider transport ecosystem that they will operate within, could be achieved in a number of ways, including attacks described in this chapter, as well as more advanced attacks that target weaknesses in new technologies such as those based on artificial intelligence. For example, researchers have demonstrated attacks that "poison" the training data sets used by machine learning algorithms to cause the system deploying those algorithms to learn based on biased data [2.27]. Other attacks have been demonstrated that exploit the insufficient specification of object classification algorithms causing vehicle perception systems to incorrectly classify objects and initiate an unintended maneuver [2.28].

Addressing Cybersecurity—People, Process, and Technology

The modern vehicle with increasing levels of intelligence, connectivity, and automation can be considered to be a cyber-physical system of systems. The connected vehicle itself is part of a larger smart mobility system of systems, for which no single organization has overall responsibility, and should therefore be assumed to have emergent properties. This is exacerbated by the need for systems to be updated in the field, for example, by OTA software update; evolving system configurations mean that the complete behavior of the integrated smart mobility system of systems throughout its life cannot be fully predicted when any of its constituent systems are first developed.

This complexity presents challenges for all aspects of dependable systems, including functional safety, functional performance, and cybersecurity. Given the long lifetime of vehicles and the wider connected infrastructure, the rapid pace of technological change, and the adaptive nature of human threat actors, the threat landscape faced by vehicles will evolve significantly over their lifetime.

Due to this dynamic and human element, it is not possible to justifiably claim that any system is 100% secure. Therefore, a comprehensive cybersecurity engineering approach must be adopted, which involves building a set of layered cybersecurity measures aimed at preventing, detecting, and responding to attacks. Such an approach ensures that vehicle systems are designed to be resilient and can maintain the required dependability when operating in the wider system of systems over their lifetime, even when faced with evolving threats. To develop and maintain this resilience, cybersecurity engineering must start from the very beginning of the product lifecycle and embed a comprehensive set of measures involving people, process, and technology in all stages: from the initial concept through product development, production, operations, maintenance, and decommissioning.

As well as technical cybersecurity activities, organizations should establish cybersecurity governance and management systems. Such a holistic approach is critical as the cybersecurity of the vehicle and its connected environment could be catastrophically compromised by only a single weak link.

Management of Cybersecurity

Management of cybersecurity is a key aspect to ensure that the technical aspects of cybersecurity are considered within an appropriate governance framework. To be effective, all cybersecurity engineering activities should be managed and coordinated effectively, compliant with applicable legislation; subject to ethical approaches to research, investigation, and vulnerability disclosure; and protected using appropriate levels of security (both cyber and physical).

Important aspects to consider in the management of cybersecurity are:

- Top management commitment.
- Cybersecurity processes.

- Cybersecurity culture.
- Cybersecurity awareness and competence.
- Continuous improvement.
- Information sharing.

In order to properly carry out the necessary cybersecurity activities and maintain a positive cybersecurity culture, it is essential that the organization has suitably qualified and experienced personnel-fulfilling cybersecurity-relevant roles throughout the organization.

As part of a cybersecurity culture, organizations also need to establish mechanisms for sharing cybersecurity-related information both internally and with external entities, such as suppliers, customers, government, and appropriate industry groups such as Information Sharing and Analysis Centers (ISACs).

To ensure appropriate and consistent effort is directed to cybersecurity activities throughout the lifecycle of vehicles, organizations need to establish policies, processes, and methods to be applied by their personnel. These should cover all cybersecurity activities during product definition, development, production, operations, and decommissioning to ensure that cybersecurity is appropriately considered throughout.

Cybersecurity Engineering

In order to address threats that are known or reasonably predictable during the development of vehicles and components, it is essential to undertake a proactive cybersecurity engineering approach to the development. A systematic security-by-design approach will not only deliver an initial level of robustness and resilience to attack but will also provide a firmer base on which further cybersecurity measures can be implemented as necessary during the operational phase of the vehicle lifecycle.

Current state-of-the-art proactive cybersecurity engineering involves a range of analysis and requirements management activities during the design phase based on a risk-based systems engineering approach:

- Threat analysis and risk assessment.
- Cybersecurity concept development and elicitation of cybersecurity requirements.
- Specification, design, and implementation of cybersecurity controls.
- Cybersecurity design analysis and verification.
- Integration verification and validation.
- Cybersecurity audit and assessment.

The proactive cybersecurity engineering approach can only address the threats associated with known vulnerabilities, or with previously unknown vulnerabilities that can be identified through verification and validation activities. However, given

FIGURE 2.3 Proactive and reactive cybersecurity engineering and the corresponding ISO/SAE 21434 clauses.

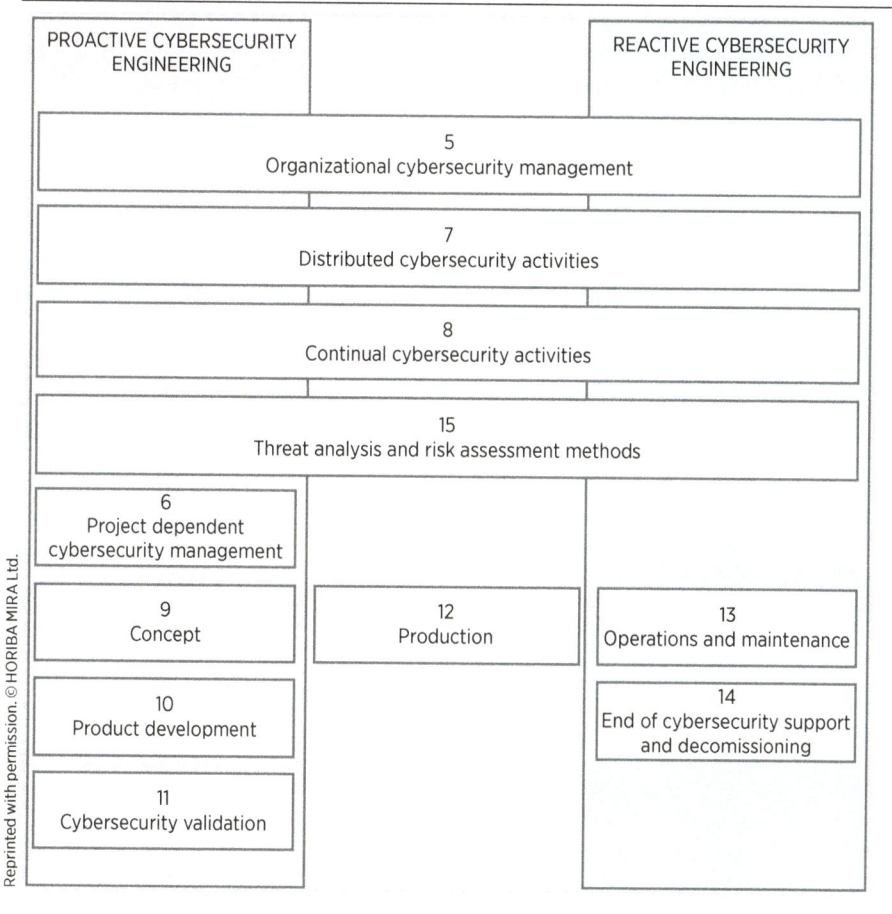

PROACTIVE CYBERSECURITY ENGINEERING		REACTIVE CYBERSECURITY ENGINEERING
5 Organizational cybersecurity management		
7 Distributed cybersecurity activities		
8 Continual cybersecurity activities		
15 Threat analysis and risk assessment methods		
6 Project dependent cybersecurity management		
9 Concept	12 Production	13 Operations and maintenance
10 Product development		14 End of cybersecurity support and decomissioning
11 Cybersecurity validation		

Reprinted with permission. © HORIBA MIRA Ltd.

the long lifecycle of vehicles, the increasingly rapid pace of technological change, and the ingenuity of attackers, it is certain that new threats will emerge that could not be foreseen during the initial product development.

Consequently, provision must also be made for reactive cybersecurity engineering approaches that are based on ongoing monitoring, detection, and response to emerging threats that are identified during the operational phase of the lifecycle.

In light of the above, regulations, standards, and best practices for automotive cybersecurity are all espousing a combination of proactive and reactive approaches to cybersecurity engineering. These two approaches can be mapped to the cybersecurity engineering activities in ISO/SAE 21434 as shown in Figure 2.3 and are discussed in more detail in Chapter 3.

Skills Required for Cybersecurity

To address cybersecurity effectively a range of multidisciplinary skills are needed. It is important to recognize that cybersecurity is not a purely technical discipline but requires competence in other areas including management, legal, human factors, and public relations. Technical expertise is required in terms of domain knowledge related to the different vehicle and mobility functions and the underlying technologies; the cybersecurity aspects of those technologies, including typical weaknesses, vulnerabilities, attacks, test methods to assess susceptibility to the vulnerabilities, and appropriate countermeasures against the attacks.

Engineering skills include systems engineering, software development, embedded systems hardware, cryptography, wired, and wireless communications. Expertise in closely related disciplines such as safety and privacy are also key. Important soft skills include creativity, tenacity, and lateral thinking and the ability for practitioners to "think like the attacker" to identify new vulnerabilities and understand how systems might be attacked.

Scientific research skills are also essential to increase the understanding of emerging technologies and associated threats and to develop novel ways of applying those technologies to more effectively manage risk in a continuously evolving threat landscape. Cybersecurity is a relatively young discipline and currently lacks a scientific basis for many of the techniques currently in use [2.29].

Technology

Technical solutions to address cybersecurity threats and achieve security requirements are now becoming widely available in the automotive industry. For example, secure communications standards are under development by IEEE and ETSI to provide security and privacy for vehicle-to-vehicle and vehicle-to-infrastructure communications. Within the vehicle, authenticity, integrity, and confidentiality of data on the in-vehicle network can be achieved through the use of cryptographic services, protocols, and algorithms, which are increasingly supported by modern microcontrollers and embedded software. Such methods are also now defined in common industry frameworks, such as the AUTOSAR Secure On-board Communications specification [2.30]. The integrity and robustness of the embedded systems themselves are also important and can be realized through hardware platforms that provide hardware security modules [2.31, 2.32]. These hardware solutions can be used to realize cybersecurity services such as secure storage, trusted boot, secure software updates, and tamper resistance features. These features are widely supported in new microcontroller and SoC products and are now also defined in the industry recommended practices [2.33].

Despite a range of security technologies and solutions being available, recall from "Challenges" (see Section 2.2) that we cannot justifiably claim to have achieved 100% security even when implementing all possible solutions. Attempting to do so is also infeasible from a cost perspective. Therefore, it is necessary to use a risk-based approach combined with systems engineering, which we cover in more detail in Section 3.5, to select appropriate cybersecurity controls that are expected to appropriately treat the identified risks. The specified controls should implement a combination of prevention, detection, and response measures as part of an overall cybersecurity concept. In addition, it is also important to establish operational processes, as described in Section 3.7, to detect, understand, and react to incidents and emerging threats throughout the lifetime of the vehicle.

Continuous research and development into new and improved attacks and cybersecurity controls is critical to enable cybersecurity risks to be continuously addressed. Technological solutions can assist here in the form of automation and intelligence to aid the ongoing search for vulnerabilities in vehicle systems, for example, techniques based on artificial intelligence are being proposed to identify weaknesses and vulnerabilities in software [2.34], as well as to automate the detection of abnormal conditions in vehicle networks and components [2.35].

Final Thoughts

Enough of the challenges, so where do we go from here? In the next chapter, we describe how to establish a cybersecurity process as a means of having a managed and structured approach to addressing cybersecurity.

3

Establishing a Cybersecurity Process

The foregoing chapters have identified the need for addressing cybersecurity in the automotive context and some of the specific challenges. Standards such as ISO/SAE 21434 are based on a process framework in order to have a managed and structured approach to cybersecurity. We now explore this subject further.

General Aspects of a Cybersecurity Process

A typical dictionary definition of a process is "a series of actions or steps taken in order to achieve a particular end." Standards such as ISO/SAE 21434 define a process; specifically in ISO/SAE 21434, the actions and steps given are undertaken in order to manage cybersecurity risk and to ensure that assets are sufficiently protected against threat scenarios.

Such standards often present a process model, identifying the activities and characteristics that are required in a conforming process. However, the intention is often that the users of the standard should integrate the requirements of the process model into their own processes. Sometimes the processes given can be adopted directly (e.g., the software lifecycle in ISO 26262 Part 6), but a significant characteristic of ISO/SAE 21434 is that a definitive lifecycle is not given instead an example workflow based on the classical systems engineering "V" model is given along with templates for a typical design phase and a typical integration phase. Thus adopters of the standard are required to adapt it into their own activities.

The approach to adapting the standard is referred to as "tailoring." "Tailoring" is defined in ISO/SAE 21434 as "omit[ting] or perform[ing] an activity in a different manner compared to its description in this [standard]." Where tailoring is applied, then a rationale is required that the adapted process is sufficient and adequate to meet the objectives of the standard. Some practical observations on tailoring include:

- Although not specifically mentioned in ISO/SAE 21434, tailoring can also include the adaptation of the standard into the specific processes or procedures within an organization. This is analogous to "project-independent lifecycle tailoring" found in ISO 26262 Part 2 Clause 5.

- Typical use cases specifically mentioned in ISO/SAE 21434, and analogous to the project-specific lifecycle tailoring found in ISO 26262 Part 2 Clause 6, include reuse, developing a component out-of-context, utilizing an off-the-shelf component, or updating a product.

- Tailoring does not apply to distributed development, these activities undertaken, for instance, by a supplier are not tailored "out" although some joint activity on tailoring between a customer and a supplier may be undertaken.

- A specific topic mentioned in the standard is the use of alternative lifecycle approaches such as "Agile." While the standard does not state this, such alternatives also must be viewed as a "tailoring" and therefore supported by a rationale particularly in the case of omitted activities. In particular, it needs to be ensured that lifecycle models such as "Agile" do not undermine the required integrity of the development process.

Standards and Best Practice

Definitions of processes are usually found in standards and related documents. Broadly speaking, documents fall into three categories as shown in Figure 3.1.

Specifically considering the role of standards, these are usually written from an engineering perspective and have the motivation of defining a common approach and establishing a "common language" among stakeholders, including the supply chain and regulators.

Standards frequently specify "what" is needed but not "how"—the latter is often closely related to IP (and in the context of cybersecurity the "how" can be related to information that is useful to malicious actors).

Guideline documents such as MISRA C [3.1] often "fill in the gaps" and provide user-oriented guidance.

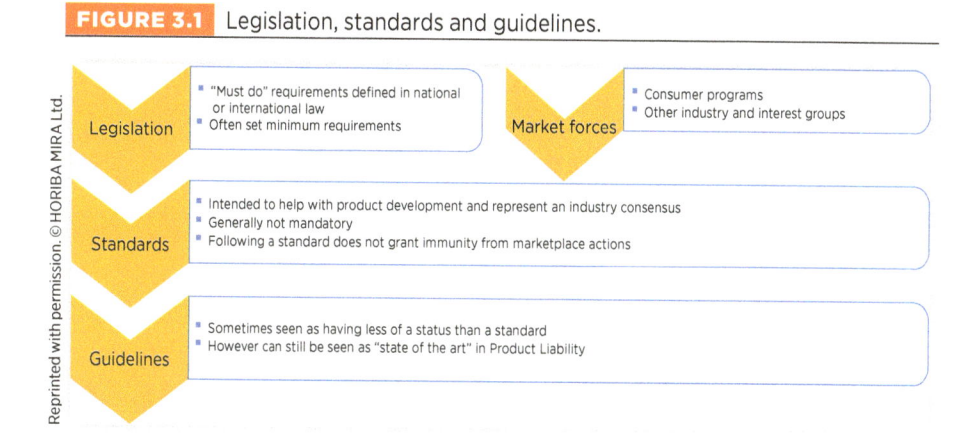

FIGURE 3.1 Legislation, standards and guidelines.

- Legislation
  - "Must do" requirements defined in national or international law
  - Often set minimum requirements
- Market forces
  - Consumer programs
  - Other industry and interest groups
- Standards
  - Intended to help with product development and represent an industry consensus
  - Generally not mandatory
  - Following a standard does not grant immunity from marketplace actions
- Guidelines
  - Sometimes seen as having less of a status than a standard
  - However can still be seen as "state of the art" in Product Liability

Reprinted with permission. © HORIBA MIRA Ltd.

Note that where specific legislation and/or standards do not exist, "catch all" legislative requirements concerned with product liability and product safety generally require compliance with the "state of the art" and "best practice" at the time of bringing a product to market. Generally, therefore, specific legislation represents the baseline or minimum set of requirements, standards represent "good practice" and "state of the art," and guidelines supplement standards represent "best practice" and may also contribute to the "state of the art."

See Section 4.2 for a summary of the interaction between legislation and standards specifically in automotive cybersecurity.

Cybersecurity Lifecycle

Cybersecurity is a continuously developing discipline, affecting products and systems from their initial design and throughout their operational lifetime. This demands an approach to cybersecurity that extends throughout the whole lifecycle—a requirement recognized in the new UN regulation and ISO/SAE 21434.

The systems engineering lifecycle V-model is well established and used by other engineering disciplines such as functional safety, with ISO 26262 specifying activities based on the V-model. ISO/SAE 21434 follows a similar approach with lifecycle phases that align with those in ISO 26262; however, it also acknowledges the dynamic nature of cybersecurity risk by incorporating an iterative risk management approach at its core. This results in an alternative view of the lifecycle with a cyclic nature, as represented in the Q-shaped image that appears in ISO/SAE 21434 as shown in Figure 3.2.

This new lifecycle view and the requirements in the standard provide more flexibility to users of the standard to tailor the activities to different process models, for example, agile development in which development progresses iteratively through a series of sprints.

While recognizing these alternative lifecycle models, the standard gives the example of a cybersecurity engineering process based on the V-model, with activities defined covering the concept phase and product development at different architectural levels, which will be familiar to those already using ISO 26262. In ISO/SAE 21434

FIGURE 3.2 Cyclic lifecycle model in ISO/SAE 21434.

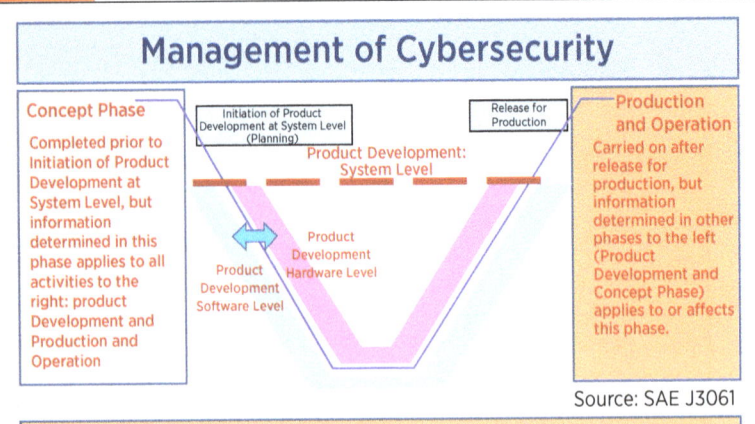

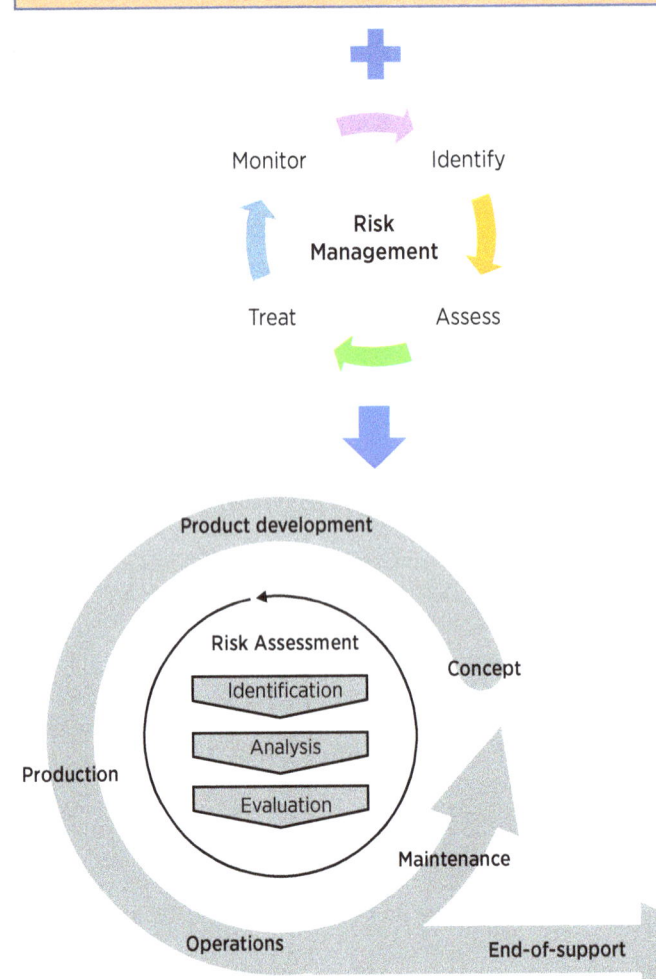

Automotive Cybersecurity: An Introduction to ISO/SAE 21434   39

the product development activities are not explicitly separated into the system, hardware, and software levels; however, an example is given as to how the cybersecurity activities during product development can be iterated at two levels: component and subcomponent, as shown in Figure 3.3.

This provides opportunities to align the cybersecurity activities to those of related disciplines (which the standard also requires in Clause 5), such as functional safety and SOTIF, in an aligned lifecycle model as illustrated in Figure 3.4. This enables organizations to establish an integrated engineering process that incorporates the needs of all related disciplines, as well as ensuring that activities that influence each other are synchronized, for example, aligning risk assessment activities of functional safety and cybersecurity enables efficient identification of cybersecurity threats that

FIGURE 3.3 Example of iteration of cybersecurity activities in ISO/SAE 21434.

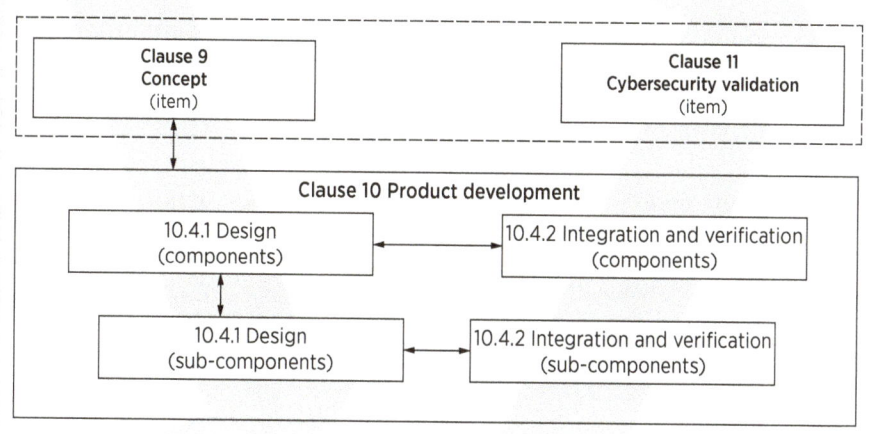

FIGURE 3.4 Aligned lifecycle combining cybersecurity, functional safety and SOTIF.

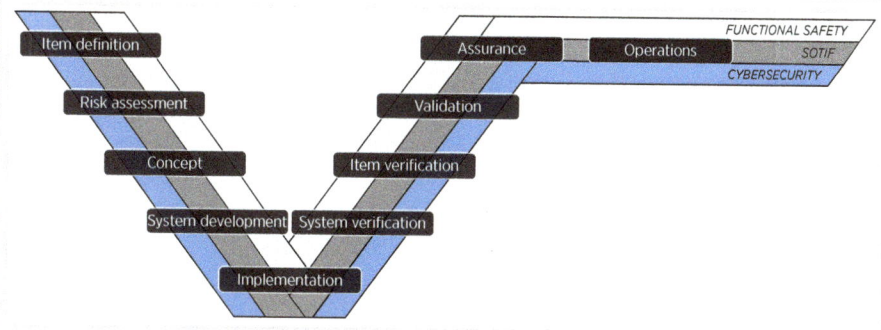

can lead to hazards and consistent rating of the safety impact of cybersecurity damage scenarios. Similarly, conflicting safety and cybersecurity requirements can be identified and resolved, as well as identifying common controls that can be used for both functional safety and cybersecurity risk reduction.

Management of Cybersecurity

Cybersecurity management takes many forms as described in the following sections.

Top Management Commitment

At the root of an organization's ability to properly address and manage cybersecurity is the commitment from the highest level of management within the organization that the appropriate priority will be given to manage cybersecurity risks. This should come from an accountable C-level executive, for example, a CISO or CTO to ensure that the commitment is made on behalf of the whole organization. The commitment can take the form of a cybersecurity policy document that acknowledges cybersecurity risks related to the organization's products and services. Such a policy may also place cybersecurity in the context of the organization's other commitments, policies, and objectives.

Cybersecurity Processes

The organization's approach to cybersecurity as defined in the cybersecurity policy should be enforced by a set of more detailed cybersecurity processes, which cover cybersecurity activities during the complete vehicle lifecycle. Cybersecurity processes may include engineering procedures, methods, design rules, guidelines, and templates. The cybersecurity processes typically specify how the requirements of cybersecurity standards (such as ISO/SAE 21434) and regulations are to be realized through the organization's own processes and how cybersecurity activities are integrated into the overall management, product development, and operational processes.

Cybersecurity Culture

A positive cybersecurity culture should be established and maintained in the organization, which promotes awareness of cybersecurity and its priority as a quality attribute of the organization's products and services. A positive cybersecurity culture is characterized by several indicators, including motivating and incentivizing staff to take a proactive attitude to cybersecurity and penalizing those who take steps that could jeopardize cybersecurity. Other indicators of a strong cybersecurity culture include mechanisms for sharing information related to cybersecurity and encourage reporting of cybersecurity concerns and incidents so that they can be rectified, improvements made, and lessons learned. Cybersecurity culture is generally difficult to capture formally in policies and processes, but may rather be something that a visitor to the organization, for example, an auditor or assessor, witnesses by speaking to the organization's personnel.

Roles and Responsibilities

ISO/SAE 21434 requires responsibilities for cybersecurity activities to be assigned to members of the organization. The standard does not mandate specific role titles or organizational structures, but allows organizations the flexibility to implement the most appropriate role structures to achieve this.

When determining the appropriate organizational structure and roles for cybersecurity, a number of key objectives should be kept in mind. A member of the board-level management of the organization should have overall responsibility for ensuring that cybersecurity is managed by the organization. In addition, there should be sufficient management, technical, and other roles defined to be able to carry out the cybersecurity activities across the organization's products, and with the appropriate level of authority to discharge those responsibilities.

For some functions, such as cybersecurity assessors, auditors, verification, and validation, it is appropriate to allocate roles that have a level of independence from the engineering teams developing the products. For example, cybersecurity auditors or assessors may belong to a dedicated quality assurance or other impartial function. Testing activities to support cybersecurity verification or validation, such as red-teaming or penetration testing, may be assigned to a dedicated department responsible for identifying weaknesses and vulnerabilities in the organization's products, without the often-conflicting objective of meeting the development program milestones.

Cybersecurity Awareness and Competence

As well as assigning resources to the cybersecurity activities, it is important to ensure those resources have an appropriate level of awareness and competence in cybersecurity to carry out their responsibilities. Appropriate competence can range from a general level of cybersecurity awareness for all staff in the organization to specific technical competencies, including, but not limited to, cybersecurity risk management, cybersecurity testing, cryptography, audit, and assessment.

Cybersecurity requires a multi-disciplinary approach including both technical and socioeconomic competencies. Applicable technical competencies include knowledge of different vehicle and mobility technologies and their cybersecurity aspects, including the underlying technologies, relevant vulnerabilities and attacks, test methods to identify and analyze vulnerabilities, and mitigation measures and engineering skills in the areas of vehicle electronics, embedded software, semiconductor hardware, IT security, network communications, data science, and simulation. In addition, competencies beyond purely technical ones are also required to cover areas such as human factors, user experience, legislation, and public relations—reflecting the human dimension to cybersecurity.

Implementing a program of coordinated training and development activities ensures that the organization has the right competencies at the appropriate levels to carry out the cybersecurity activities effectively. These activities can include formal education and training courses in specific aspects of cybersecurity, attendance at conferences, and other events to maintain an up-to-date awareness of the latest threats, attacks, and mitigation approaches to on-the-job development through coaching and shadowing more experienced team members.

Continuous Improvement

A key element of any process is monitoring its ongoing use and effectiveness, capturing lessons learned during application of the process on real projects, and using those lessons to improve the processes. This is particularly important for a cybersecurity process since, as well as the fact that the threat landscape is continuously changing, cybersecurity is still a nascent engineering discipline with many opportunities to improve the processes and methods based on shared learning from across the organization and even from other industry sectors. Therefore any cybersecurity engineering process should have a mechanism for regular review and a means to gather feedback from the users of the process and incorporate that in future versions of the process.

Information Sharing

Information sharing is important for cybersecurity as it encourages collaboration between different parts of an organization, and between different organizations, helping to prevent, detect, and understand threats, their impacts, and how they can be mitigated effectively. Information sharing can include exchanging of knowledge about threats, vulnerabilities, mitigations, best practices, or tools relevant to the organization's products.

Sharing information between organizations who manufacture products is often limited due to concerns over disclosure of intellectual property anticompetitive concerns. However, the motivation for sharing appropriate cybersecurity information is that a threat to one organization's products can also be a threat to similar products from other organizations. This motivation is the driver for establishing industry information-sharing initiatives such as the Auto-ISAC [3.2], which has members from across the automotive supply chain and facilitates information sharing in a trusted environment for the collective benefit of the members.

Proactive Cybersecurity Engineering

A critical component of cybersecurity engineering for all vehicles or components is a systematic security-by-design approach. Considering cybersecurity starting from the initial stages of development will not only increase confidence in the level of robustness and resilience to attack when the product is launched but also lays the foundations for vehicles to adapt to emerging threats during the operational phase of the lifecycle.

The new UN Regulation 155 requires vehicle manufacturers to demonstrate that they have implemented effective proactive cybersecurity engineering in the form of risk assessments, implementation of appropriate mitigations, and testing their effectiveness. This reflects the approach in other best practices including ISO/SAE 21434, the Auto-ISAC best practices, and the NHTSA cybersecurity guidelines [3.3].

ISO/SAE 21434 specifies the following activities that together constitute proactive cybersecurity engineering:

Cybersecurity Responsibilities at Project Level

For individual development projects, a project team structure such as the example in **Figure 3.5** should be established to define the responsibilities for the specific cybersecurity activities required for that project. This example project structure has the following key elements:

- A project should have an overall "cybersecurity manager" responsible for ensuring the cybersecurity activities are carried out for the project. This can be a specialist from a dedicated cybersecurity team or the project team or even the project leader.

- The cybersecurity manager would not necessarily carry out all the cybersecurity activities on a project, but rather members of the development team and other subject matter experts should participate in cybersecurity activities, such as

FIGURE 3.5 Example of allocation of responsibilities for cybersecurity activities.

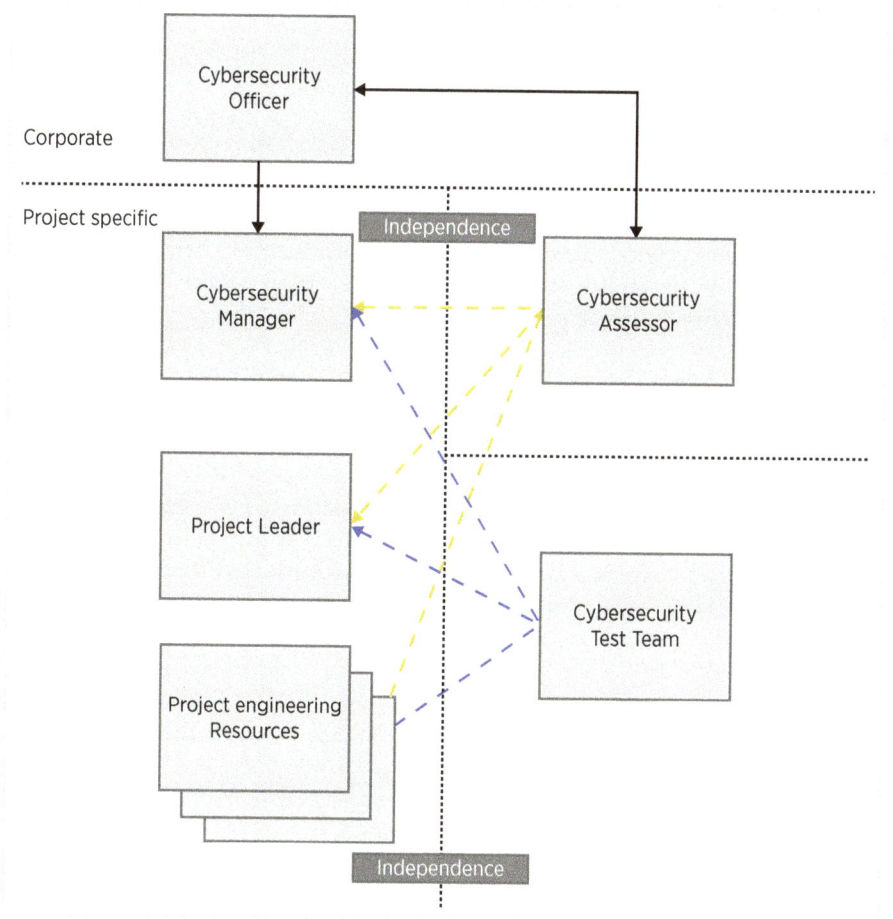

carrying out threat analysis and risk assessments; generate and review cybersecurity requirements; and prepare cybersecurity verification and validation plans.

- Some cybersecurity activities on a project should be carried out with a level of independence from the development team, for example, cybersecurity testing can include testing carried out by an independent quality assurance team or penetration testing by a dedicated red team, whose objectives are to try and find weaknesses in the vehicle or its components.

- Cybersecurity assessment should also be carried out with a level of independence from the development team, with assessors from either another department of the organization or a separate organization. This maximizes the impartiality of the assessment of whether the cybersecurity objectives for the project are achieved.

Cybersecurity Planning

Before detailed planning for the cybersecurity activities starts, ISO/SAE 21434 requires that the cybersecurity relevance is determined for the item or component being developed. This means considering whether the item or component can be exposed to threat scenarios that can result in damage to a stakeholder. This relevance assessment helps to ensure that the efforts required for cybersecurity engineering activities are applied appropriately and to those developments that require it, but not to those items or components for which cybersecurity is not a concern. Any development that is determined not to be cybersecurity relevant is then not subject to further cybersecurity activities and can be developed according to a general quality-managed engineering process.

Cybersecurity relevance can be determined in a number of ways, including using a specific checklist or flowchart (an example appears in ISO/SAE 21434 Annex D) or by using the organization's prior experience from similar products. The principle of the example flowchart in ISO/SAE 21434 is that any electrical/electronic item or component that contributes to safe operation processes personal data of the occupants or otherwise uses network components that are considered cybersecurity relevant. Note that, in practice, this means that any ECU connected on an in-vehicle network is cybersecurity relevant and should be developed in accordance with ISO/SAE 21434.

In order to clearly define and manage the cybersecurity engineering activities for a specific development project and to enable efficient and effective use of the assigned responsibilities, a cybersecurity plan should be developed and maintained throughout the project.

As well as providing typical project plan content, such as a breakdown of the cybersecurity activities, assigned resources to complete the activities, and timing information, it is important to identify any dependencies on other activities (including those that are not cybersecurity specific) that may produce inputs to the cybersecurity activities. The cybersecurity plan should also identify the work products that each activity will produce, taking into account the classification from an information security point of view. This ensures that the relevant individuals have access to the work products but that sensitive information is treated on a "least-privilege" basis.

The cybersecurity activities are, of course, not planned in isolation, and there may be other plans relating to other aspects of a development project. As such, the cybersecurity plan can be integrated with a safety plan (in accordance with ISO 26262) or even included in a larger project plan. In any case, the overall project plan must reference the cybersecurity plan and associated activities.

The cybersecurity plan should be maintained and refined incrementally during development. Any changes to the project activities, including task refinements, changes to assigned individuals, or timing changes, should be updated in the cybersecurity plan.

Den Rise/Shutterstock.com.

ISO/SAE 21434 requires that the cybersecurity plan is maintained until the release for post-development. After development, the standard requires other types of plans to be generated to document cybersecurity activities in other lifecycle phases, for example, a production control plan and incident response plans relating to specific incidents. It may be helpful to consider how the cybersecurity plan relates to other such plans, the key point being to ensure that no gaps exist in documenting how cybersecurity activities in any lifecycle phase will be carried out.

For some development projects, the cybersecurity activities may need to be tailored, for example, if an item or component is being reused, if a component is being developed out of context, or if an off-the-shelf component is being integrated. In each of these scenarios, some of the cybersecurity activities may need to be modified or omitted. Any such changes to the process should be made with careful consideration to whether the objectives of ISO/SAE 21434 can still be achieved and do not compromise the overall cybersecurity engineering of the product being developed. Therefore any tailored activities need to be documented with rationales in the cybersecurity plan.

See Section 4.1.5 for further discussion of the use of "off the shelf" or "out of context" products and the additional evidence or activities that an integrator may need to undertake.

Concept Phase

The concept phase is the first phase of the product lifecycle covering the initial cybersecurity activities and should be aligned with the equivalent functional safety concept phase activities specified in ISO 26262. One of the objectives of the concept phase as defined in ISO/SAE 21434 is to determine a set of cybersecurity goals using a risk-driven approach. A cybersecurity concept is then developed, which specifies implementation-independent cybersecurity requirements and the strategy to achieve the cybersecurity goals. The following activities are carried out in the concept phase.

Item Definition

To support the risk-based approach to developing the cybersecurity concept, the first activity in the concept phase is to create a description of the item being developed. The item is the ISO/SAE 21434 term for the component or set of components being developed that provide a function at the vehicle level. This is in alignment with the same term used by ISO 26262 for functional safety, and indeed a single item definition document containing a superset of cybersecurity and functional safety-related information can serve both disciplines.

The item definition describes the vehicle feature being developed and should include a description of the main functions and behavior defined during each lifecycle phase, that is, the functionality required not only during the operational phase but also during production and at decommissioning for example. The preliminary architecture of the item is described, as well as the boundary of the item so that the item can be distinguished from its operational environment. The operational environment can include other non-vehicle aspects that may be relevant when considering cyber-attacks and may also include other items with which the item being developed interacts. For some items, the cybersecurity item boundary may be different from the functional safety item boundary, for example, an element that is not safety related might be included in the cybersecurity item, but not the functional safety item. The item definition should also document any assumptions made, including those regarding any external systems or the operational environment.

For components being developed out of context (as described in Section 3.5.2), for example, an ECU, microcontroller, or software component, the vehicle-level functionality may not be known. In this case, information about the item can be assumed, perhaps based on information or requirements from potential customers of the component, and the assumed information documented in an item definition. These assumptions should be validated at a later stage of development, for example, during distributed development discussions during procurement of a component.

Threat Analysis and Risk Assessment

Threat analysis and risk assessment is a central element of the cybersecurity engineering process, whose purpose is to identify, analyze, and treat cybersecurity risks relevant to the system. Although this activity is first carried out during the concept phase, it is essential that it is carried out iteratively throughout development and beyond to ensure that those threats that can be identified at the design stage are assessed for the associated risk and treated to ensure the risk is at an acceptable level. The following activities are required for effective threat analysis and risk assessment:

- Develop architectural and functional models of the system to be analyzed.
- Identify cybersecurity relevant assets and their cybersecurity properties (e.g., confidentiality, integrity, and availability).
- Identify possible damage scenarios and threat scenarios associated with the assets.
- Rate the impact of the identified damage scenarios in the relevant impact categories (safety, financial, operational, and privacy).
- Identify and/or update the attack paths that an attack could use to realize a threat scenario.
- Assess the ease with which identified attack paths can be exploited.
- Determine the risk value of each threat scenario and assess its acceptability.
- Determine the appropriate risk treatment options.

In order to identify threat scenarios and attack paths, it is necessary to develop a threat model of the system under consideration. A threat model takes a system-centric view of cybersecurity risk, based on analyzing interactions between elements of the system and how they may be attacked. There are various methods for developing a threat model, for example, the STRIDE threat modelling method published by Microsoft [3.4], which can be seen as a set of guidewords to structure the identification of threats related to a system architecture. Attack trees [3.5] are a useful method of documenting attack paths by combining different steps of an attack, detailing their interdependencies, and how they may be combined to realize a threat scenario.

Various methods have been proposed for assessing the impact and attack feasibility parameters and determining cybersecurity risk. Some of these such as the Common Vulnerability Scoring System (CVSS [3.6]) are derived from the IT world and, therefore, often have limitations in their ability to reflect the particular features of cyber-physical systems such as vehicles, their safety-related functions, and their dependence on reliable environment sensor inputs.

Work carried out by the EU-supported EVITA collaborative research project [3.7] resulted in an automotive-specific method for cybersecurity risk assessment. This built upon existing standardized rating scales for impact and attack potential, adapting and combining them for the purposes of automotive cybersecurity risk. The recommendations of SAE J3061 [3.8] and the development of ISO/SAE 21434 [3.9] have made significant progress in adapting these methods further to increase their suitability for vehicle cybersecurity risk assessment.

Risk Treatment and Cybersecurity Goals

Once risks have been assessed, it is necessary for the organization to decide how to treat those risks and how that translates into cybersecurity controls that need to be implemented in the vehicle and its components, as well as in off-board systems, such as servers and mobile apps. It is also important to understand which aspects cannot be resolved under the control of the vehicle manufacturer or its supply chain and, therefore, what assumptions must be made and validated in the operational environment. This includes risks that may be due to threats that are either launched from or targeted against off-board systems.

To determine an appropriate set of cybersecurity controls, the identified and assessed risks should first be evaluated against the level of acceptable risk, which must be determined by an organization based on various factors. The acceptable risk can be determined based on thresholds that may be imposed by applicable legislation or regulations, the nature of the products or services the organization is providing, and consideration of established moral or societal factors.

Any risks that are above the acceptable risk threshold must be treated. The first step is to determine from a set of possible options the approach by which a risk should be treated; risk treatment options can include avoiding the risk by redesigning the system to remove the source of the risk; reducing the risk by implementing a control to reduce its likelihood or impact; transferring the risk to another party, for example, through an insurance policy; and accepting or retaining the risk if it is sufficiently low or a rationale can be provided as to why the risk can be retained.

With the approach to risk treatment determined for each risk, a set of cybersecurity goals can be defined for each of the risks that are to be treated by reducing the risk. Cybersecurity goals are high-level objectives that form the top level of a subsequent hierarchy of cybersecurity requirements, as part of a systems engineering approach to specifying, designing, implementing, verifying, and validating suitable cybersecurity controls to address the risks.

CAL

CAL is intended to be the cybersecurity equivalent of the ASIL in ISO 26262 and provides a means to appropriately scale the effort and rigor required by the later cybersecurity engineering process activities. The motivation is to provide a justifiable means of determining "how much is enough" in order to appropriately focus engineering efforts and avoid overengineering. Annex E of ISO/SAE 21434 describes an informative approach for determining and using CALs.

ISO/SAE 21434 states that a CAL should be determined for each threat scenario based on the associated impact and attack vector. This approach appears similar to the determination of risk values; however, it should be noted that while the risk value is dynamic, the CAL is intended to remain stable during development since it forms part of a development requirement. Therefore the CAL should be determined based on the less dynamic factors that constitute risks. This is the reason why ISO/SAE 21434 Annex E suggests using impact in combination with the attack vector approach to determine a CAL, rather than any of the other attack feasibility rating approaches; the rationale is that the attack vector approach looks only at the aspects of architecture that cause a system to be exposed to a cyber-attack, rather than the aspects of the attack itself.

The CAL can then be used to specify the rigor necessary for the subsequent engineering process (in a similar way to the ISO 26262 ASIL) by assigning the CAL as an attribute of the cybersecurity goal that has been specified to address the corresponding threat scenario.

ISO/SAE 21434 includes some examples of how the rigor of some cybersecurity development activities can be scaled based on the CAL. Activities that can be scaled based on CAL include:

- The extent and scope of design verification methods.
- The extent and parameters for cybersecurity verification and validation methods, such as fuzz testing and penetration testing.
- The level of independence needed for cybersecurity verification, validation, and assessment activities.

In principle, the CAL concept can be extended to other cybersecurity activities for which scaling of the depth or extent of the activity is appropriate. There is a considerable ongoing debate within the industry as to whether more specific assignments of development methods and measures for different CAL can be established, and at the time of writing, further projects are ongoing within ISO and SAE to explore this.

One possible direction is to use the CAL to specify a required level of resistance to attack that the cybersecurity controls need to be designed to withstand. This idea is seen in the similar concept of Evaluation Assurance Levels (EAL) from the Common Criteria [3.10], in which the EAL specifies the level of attack potential (i.e., capabilities of an attacker) that will be considered during a security evaluation of a product. A possible scheme for specifying the attack potential that an item or component should resist based on CAL is shown in Table 3.1.

TABLE 3.1 Use of CAL to specify required attack resistance.

CAL	Required assurance	Required attack resistance
CAL1	Low to moderate	Basic
CAL2	Moderate	Enhanced-basic
CAL3	Moderate to high	Moderate
CAL4	High	High

Cybersecurity Requirements and Controls

A systems engineering approach should be used to specify requirements for the system design to achieve the cybersecurity goals. These cybersecurity requirements should be treated as any other engineering requirement; this includes integrating and managing cybersecurity requirements together with those related to other engineering disciplines such as functional safety requirements, functional requirements, and performance requirements. This enables any conflicts between requirements of different disciplines to be identified and resolved and any synergies between the needs of the different disciplines to be exploited, for example, through the specification of controls that fulfill both safety and security requirements.

During the different product development phases, cybersecurity requirements are refined at different architectural levels, from the cybersecurity goals at the vehicle level to cybersecurity requirements at the hardware and software levels. Unlike ISO 26262, ISO/SAE 21434 does not define specific names for each level of requirements, but instead provides flexibility for organizations to define the levels of cybersecurity requirements as required. The following example cybersecurity requirements hierarchy could be adopted and is illustrated in Figure 3.6:

- **Cybersecurity goals**—high-level objective to mitigate the risk of the associated threat scenario.
- **Concept-level cybersecurity requirements**—design-independent strategy to achieve a cybersecurity goal; may be aligned to preventing threat scenarios and protecting cybersecurity properties of assets.
- **System-level cybersecurity requirements**—design-specific strategy to achieve one or more concept-level cybersecurity requirements; specified for the system architecture; based on vulnerability analysis of the proposed architecture.
- **Hardware and software-level cybersecurity requirements**—specified for the hardware and software implementation based on further, more detailed, vulnerability analysis.

If the CAL concept is being used, the CAL assigned to a cybersecurity goal can be inherited by all the cybersecurity requirements derived from it, enabling assurance requirements to be transferred to lower levels of the requirements hierarchy and, thus, to components throughout the supply chain. This enables

FIGURE 3.6 Example hierarchy of cybersecurity requirements.

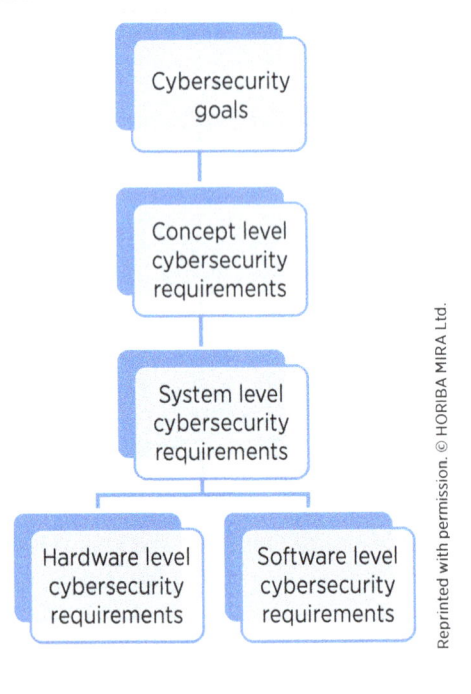

the level of assurance required by the vehicle manufacturer to be traceably converted into appropriate requirements for process rigor to be implemented by component suppliers.

Design Verification

It is important to begin verification during the design phases of the lifecycle because this allows any vulnerabilities to be discovered at an early stage, rather than waiting until the testing phases when any rework to address discovered vulnerabilities becomes increasingly expensive.

A number of design verification activities are relevant for cybersecurity, including the following:

- Requirements review.
- Design review.
- Simulation.
- Analysis by formal methods.
- Software code static analysis.
- Software code review.
- Hardware schematic review.
- Vulnerability analysis.

These activities should generally be carried out iteratively during development by all organizations within the supply chain. Executing design and verification activities in such an iterative fashion allows vulnerabilities to be discovered and fixed as early as possible in the development process.

It can be seen that a broad range of skills and capabilities are required to perform the various design verification activities at the different levels of abstraction. This includes vehicle-level knowledge and expertise in embedded systems and software, semiconductor hardware, radio frequency (RF) and radar, optics and ultrasound, as well as IT and information security.

Cybersecurity Testing

This section discusses the role of cybersecurity testing, including how it differs from other forms of testing and the different testing activities that can contribute to cybersecurity assurance.

Cybersecurity Testing Challenges

Cybersecurity threats are realized when an attacker is able to exploit vulnerabilities, which may originate from a variety of sources. Cybersecurity testing is a key method

FIGURE 3.7 Difference between intended and actual behavior.

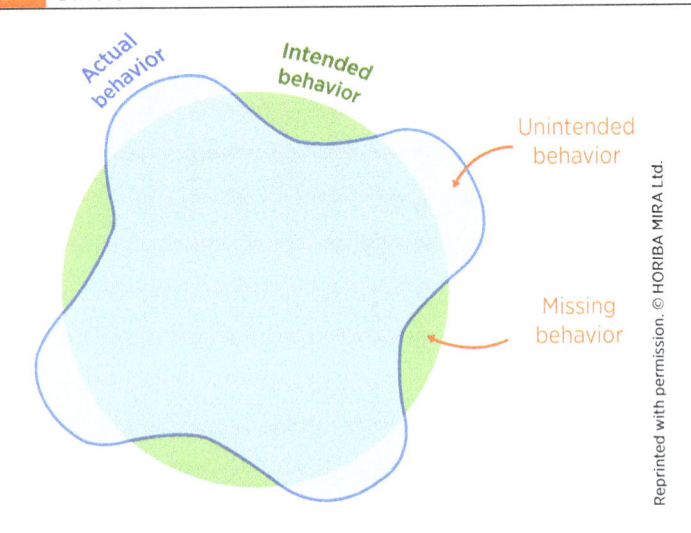

of identifying vulnerabilities, although there are several challenges [3.11] with testing for cybersecurity compared with other forms of testing.

In principle, the "implemented behavior" should be identical to the "intended behavior" that was required. In practice, however, specification flaws, design errors, and implementation defects may mean that not all of the intended behavior is actually achieved, resulting in "missing behavior," while some of the behavior that is implemented may be unwanted "unintended behavior" (see Figure 3.7). Although traditional functional testing is very good at identifying missing behavior (e.g., the system does not produce the correct outputs for given inputs), it is less capable at identifying unintended behavior since this was not a requirement, and the test plan is specifically constructed to verify compliance with the specification.

Another challenge is that we can never claim to have achieved 100% security since new threats, vulnerabilities, and attacks are continually discovered, often suddenly and with a dramatic impact on the threat landscape. An effective cybersecurity testing strategy must therefore include both testing for known vulnerabilities as well as a more exploratory search for unknown vulnerabilities.

Finally, the "coverage" metric typically used in testing is difficult to define for security due to the evolving threat landscape discussed above, and it is more appropriate to consider "assurance": grounds for justifiable confidence that an objective is achieved. Assurance may be provided through testing, as well as through design verification, process audits, or independent assessment.

Cybersecurity Testing at Different Lifecycle Phases

Testing products and services at all lifecycle phases and by all organizations within the CAV and mobility ecosystem enables verification of the cybersecurity of each individual element and also helps provide assurance that no additional vulnerabilities are introduced during integration. Cybersecurity testing is required at each of the following development phases:

- Hardware development.
- Software development.
- System integration.
- Vehicle integration.

Vulnerability analysis must also continue throughout the operational phase of the vehicle's lifecycle to identify and analyze emerging events and their impact on the vehicle and the wider system to which it is connected.

Therefore, additional activities to capture cybersecurity-relevant information during field monitoring and use that information to inform the identification of new vulnerabilities should be part of the overall engineering process.

Cybersecurity Testing Activities

Pre-testing Analysis. Prior to practical testing, threat modelling and security-focused reviews should be carried out, including security design reviews, static code analysis, source code reviews, and hardware schematic reviews. This helps to identify at an early stage any vulnerabilities that are specific to a particular design, implementation, or integration details. These activities are also important to identify potential vulnerabilities that should be investigated further for exploitability using a combination of the following practical testing methods.

Functional Testing. This involves testing whether the implementation produces the correct functional behavior under a valid set of inputs defined by the specification, for example, the correct result of a cryptographic computation. Usually, functional testing is restricted to the specified operating range and answers the question "Does the implementation deliver the intended functionality?"

As well as verifying correct response to intended inputs, it is critical to test that the target behaves correctly for out-of-specification or malformed inputs, for example, input data of incorrect length or out of the specified bounds. This is typically achieved by dynamic analysis methods such as fuzzing, which generates random or directed sets of test input data designed to exercise the system near the boundaries of its specification.

Correctness Testing. Cybersecurity-relevant functions often contain countermeasures against attacks whose correct operation is not observable through the function outputs or the interfaces of the system. Correctness testing is therefore used to verify that the internal behavior of the implementation is as expected. This activity requires alternative methods to functional testing and is usually carried out using "white box" techniques and development tools such as debuggers, simulators, or emulators.

Correctness testing activities can be carried out at the same test phase process steps as Functional Testing, although some tests may be better executed during the design phase, particularly if internal behavior needs to be monitored using an emulator.

Penetration Testing. Even extensive systematic testing for correct observable and non-observable functionality is not sufficient to test whether an implementation really resists the relevant attacks. Penetration testing is used to determine this by simulating the actions of a real attacker and, therefore, requires sufficient time and resource to adaptively follow "interesting" leads as they are uncovered.

It is also necessary to test for the presence of other vulnerabilities, for example, due to additional functionality outside the specification or due to physical characteristics of the device, such as side-channel information leakage or susceptibility to fault injection attacks. Penetration testing typically involves multi-disciplinary skills such as software, electronics, RF, and cryptography. These skills may need to be brought in from different parts of an organization or from third parties under relevant nondisclosure agreements.

Carrying out time-consuming penetration tests for all possible attacks is impractical, so a program of directed tests should be planned based on the findings of earlier activities.

Penetration testing can be carried out using a black-box approach, in which the tester puts themselves in the position of an attacker and tries to identify and exploit vulnerabilities without any prior information. An example would be injecting messages on a vehicle CAN bus without prior knowledge of the vehicle's CAN database, meaning that the tester would first need to reverse engineer the CAN database by analyzing typical messages observed during normal operation. At first glance this has the apparent advantage of being a realistic approach; however, it is not practical to test every possible attack scenario and may lead to a deep-seated vulnerability being missed due to time constraints.

A white-box approach is usually more efficient, as the tester is able to identify vulnerabilities with the help of design or implementation information. This may include design specifications or implementation details such as source code. Clearly, this approach gives the tester an advantage over a real attacker, and it is therefore important to provide a justification of how a real attacker could successfully carry out an identified attack path without access to the information. Access to up-to-date sources of threat intelligence is therefore important to assess an attacker's capabilities and motivations relative to the test conditions. However, a white-box approach does provide the opportunity to identify more vulnerabilities in the same timescale, and thus saving costs in terms of either in-house testing effort or third-party independent testing.

Due to the limited transfer of detailed design information within the automotive supply chain, in practice, a hybrid "gray box" approach may be adopted, which utilizes any available information but adopts the position of a real adversary to determine the missing details.

Vulnerability Analysis and Management

ISO/SAE 21434 defines vulnerability as a "weakness that can be exploited as part of an attack path," in a similar way to other security standards such as ISO/IEC 27000.

A weakness is further defined as a "defect or characteristic that can lead to undesirable behavior," which, in the case of a cybersecurity vulnerability, could cause damage to a stakeholder.

In order to identify vulnerabilities, it is first necessary to identify weaknesses, for which various methods can be used during design, verification, validation, and operational monitoring, for example, various review, analysis, and simulation techniques can be used during design to identify flaws in the system design, architecture, or requirements. Testing activities, such as functional testing, vulnerability scanning, and penetration testing, can be used during verification and validation to identify defects in the implementation of a system or its components.

Weaknesses must then be analyzed to understand whether they are exploitable by an attacker, and if so, how feasible are the attack paths required to exploit them. This is the role of vulnerability analysis, which makes use of techniques such as attack trees and resources such as vulnerability databases to map weaknesses to common patterns and understand how they can be leveraged in order to compromise an asset of the system.

Identified vulnerabilities then need to be managed so that they can be eliminated or mitigated by countermeasures. This involves connecting the vulnerabilities and associated attack paths with the threats that could be realized against assets and the resulting damage scenarios. Establishing these links enables the risk posed by a vulnerability to be assessed in the context of the system containing the vulnerability. An appropriate risk assessment is critical to be able to determine the most appropriate way to manage the vulnerability and thereby treat the associated risk.

This requirement to manage vulnerabilities and risks forms the basis of many of the requirements within ISO/SAE 21434, and therefore it is critical for the automotive industry to deploy practical methods to carry out these activities and provide satisfactory evidence of the outcomes to regulators, customers, and other stakeholders.

Cybersecurity During Production

The opportunity for vulnerabilities to be introduced into a vehicle or its components is not limited to the product development phase; the production environment also presents significant potential threats and vulnerabilities for exploitation by attackers. Examples of production activities that could be attacked include the provisioning of keys into ECUs at the end of line during vehicle production, storage of sensitive data in an uncontrolled production IT system, or the insertion of vulnerabilities into the product (e.g., a hardware Trojan inserted into a semiconductor component).

Cybersecurity management during production is briefly covered in ISO/SAE 21434, with a requirement to establish a cybersecurity management system (CSMS) for the production environment; however, for details of how this is to be realized, the reader is referred to different standards, such as IEC 62443, which is targeted at industrial control systems typically found in factory environments. Whichever standards or approaches are followed, it is important that an organization's CSMS for its products covers all lifecycle phases—development, production, operations, and decommissioning—and that gaps do not occur simply because separate approaches have been followed.

ISO/SAE 21434 also contains requirements to create a production control plan to ensure that any cybersecurity requirements specified during product development but that need to be implemented during production are transferred to the relevant stakeholders. Typical contents of a production control plan would be instructions for implementing the cybersecurity requirements, the specific tools and equipment required during production, and any additional cybersecurity controls needed to maintain information security or prevent unintended modification of the product during production.

It is also important to specify how any cybersecurity requirements implemented during production will be verified, for example, by tests or inspections carried out on the production line or during inspections. Finally, many production processes involve privileged access to vehicle systems, for example, end-of-line flashing of software. It should be ensured that these are carefully controlled as part of the lifecycle management of the product so that a privileged level of access intended for production personnel cannot be exploited by an attacker when the product is in the field.

Reactive Cybersecurity Engineering

The lifetime of a typical vehicle can extend to decades, much longer than many consumer electronics or other technology products, and as such, there is a significant challenge to maintaining cybersecurity over that lifetime in the face of an evolving threat landscape. Even the best-engineered vehicles, using the proactive approaches described in the previous sections, will be exposed to new threats and emergent vulnerabilities at some point during their operational life. It is therefore important to consider cybersecurity during the operational phase of the vehicle lifecycle, assuming that new and emergent cyber-attacks will occur and need to be detected and responded to.

This need is recognized in the UN Regulation 155, which contains requirements to establish processes and demonstrate the vehicle's technical capabilities for the detection and response to new threats, attacks, and vulnerabilities. This capability is not limited to detecting attacks directly on vehicles but also monitoring to maintain an awareness of changes in the threat landscape and new attack methods and vulnerabilities that may affect an organization's products. To facilitate this activity, ISO/SAE 21434 includes a range of requirements for "continual cybersecurity activities" in Clause 8, which describe how organizations should conduct monitoring, evaluate events for relevance to their products, and manage response to incidents.

Cybersecurity Monitoring

Continual monitoring of various sources of information is a key enabler for multiple cybersecurity activities during the lifecycle. Information gained during monitoring is necessary to support product development activities, such as threat analysis and risk assessment; specification of cybersecurity controls, verification, validation, planning, and development of software updates; and handling of cybersecurity incidents. The cybersecurity monitoring activity of ISO/SAE 21434 can be seen as the collection and management of cybersecurity threat intelligence, although the standard does not use this term directly. Threat intelligence includes information about all aspects of cybersecurity, including threats, vulnerabilities, and attacks, and can be collected from a diverse range of sources, for example:

- Vulnerability databases (e.g., NIST NVD [3.12], MITRE [3.13]).
- Commercial threat intelligence services.
- Supplier advisories.
- Information sharing forums (e.g., Auto-ISAC [3.2]).
- Published research in academic papers.
- Presentations at security and other relevant conferences.
- Social media.
- The organization's own internal security research programs.

Processes for monitoring should be established, which define which of these sources of information the organization will use, what information will be collected, and how it will be processed, stored, and used by the organization. With such a wide array of information in different forms, often unstructured in nature, it is critical for the organization to find ways to structure the information so it can be accessed efficiently and effectively. Therefore, the monitoring process should include methods to refine the collected information to usable knowledge, filtering out irrelevant content and focusing on establishing the significance of the information and its relevance to the organization's products.

ISO/SAE 21434 requires the definition of "triggers" to triage the collected information. The standard defines these triggers simply as criteria for triage; in practice, a trigger can take the form of a decision process that security analysts can use to establish the relevance of the information, for example:

- Is the information about a real incident that has occurred or a potential threat?
- Is the information about a specific vulnerability?
- Do any of the organization's products use the component affected?

Evaluation of Cybersecurity Events

Any cybersecurity information that is considered to be relevant to the organization's products must then be evaluated to understand its significance and priority. This activity is specified in ISO/SAE 21434 as "cybersecurity event evaluation" and aims to establish whether the event indicates a weakness in an item or component and whether a patch or "fix" is available for the weakness.

For example, a cybersecurity event may be a new vulnerability discovered in a common software component during the operational phase of the lifecycle or a new attack method published at a security conference. In such cases, it is important for vehicle and component manufacturers to understand the relevance of the discovered weakness to their products and respond appropriately.

As part of this activity, the identified weaknesses are analyzed to determine whether they could be exploited by an attacker and, therefore, considered vulnerabilities. The analyst should seek to understand whether the vulnerability is part of an existing known attack path, for example, does it appear in an existing attack tree and has the risk of the corresponding threat scenario changed? If the vulnerability is part of a new and previously unknown attack path, does that attack path lead to a known or new threat scenario and/or damage scenario? This analysis then invokes another iteration of the risk assessment, using the vulnerability analysis and management activities described in Section 3.5.8.

Any identified vulnerabilities are managed and the corresponding risks treated as part of either product development if the vehicle or component is not yet in production or incident response if the vulnerability affects a product that is already in operational use.

Detecting and Responding to Attacks

In the IT sector, security operations centers (SOCs) are used to monitor and protect enterprise information systems. SOCs can vary widely in scale and complexity, from a single analyst working alone to a dedicated facility with hundreds of staff performing different security operations roles, including event processing, deep analysis, incident response, and proactive threat hunting. SOCs typically make use of a suite of analysis and management tools, for example, a security information and event management system, which collects, collates, and correlates data from relevant sources, including:

- Analysis of network traffic.
- Network behavioral analysis.
- Logs from firewalls and antivirus systems.
- Intrusion detection and/or prevention systems.
- Cybersecurity threat intelligence.

This concept can be extended to provide equivalent functions for vehicles and transport infrastructure if the vehicles and other nodes of the network are able to provide data to the "vehicle security operations center" (VSOC). A VSOC should provide similar functionality to a conventional SOC but, in addition, needs to accommodate the broader aspects of threats faced by vehicle systems and the networks to which they are connected.

Cybersecurity Incident Response

When a cybersecurity event has been confirmed as an incident, an appropriate response is necessary. The appropriate response depends on the nature and severity of the incident, how it affects the product and its stakeholders, and may include one or more of the following:

- Actions to contain the effect or propagation of the incident.
- Deployment of software or configuration updates (e.g., by OTA software update).
- Issue of an advisory to customers or consumers.
- Recall.
- Public relations and communications activities to inform the general public.

Assessing the Effectiveness of Detection and Response

Clearly, incident response processes are put in place to handle potential undesirable scenarios with the general hope that they will never need to be used. If such processes are never invoked, then there is a risk that an organization does not have a clear view of the effectiveness of its incident response processes and how able they are to produce a timely response and the desired outcomes; for example, the organization may never know whether no incidents were raised because none actually occurred or because their detection and response processes were insufficient. It is therefore critical to regularly assess the effectiveness of monitoring, detection, and incident response processes, for example, using simulations and exercises. Exercising activities for cybersecurity incidents are analogous to the more established exercises carried out by organizations for other types of business continuity and disaster recovery planning, and suitable activities can include:

- Verification of vehicle-related threat detection and the correct understanding and response by use of the organization's processes.
- Carrying out practical tests of the detection and response capabilities by simulating attacks against test vehicles in controlled facilities and conditions, and verifying correct detection and response (e.g., vehicle penetration testing).
- Ongoing regular exercises during operation of the incident response capabilities to test their ongoing effectiveness, including processes, people, and technology.

Exercises of cybersecurity operations capabilities can be developed based on the following steps, which are based on general guidance for cybersecurity exercises by the UK National Cyber Security Centre [3.14]:

- Establish the target of the exercise, including the products, people, and processes involved.
- Gain organizational approval and commitment to the exercise.

- Determine the correct exercise format (e.g., tabletop or role play based).
- Establish the exercise team—both those who will design the exercise and those who will participate.
- Define success criteria and metrics.
- Develop the exercise scenario and simulation of events.
- Develop guidance and document templates for participants.
- Capture evidence, feedback, and lessons learned.

Updates

Updates are often discussed in the context of updating software, although replacement of hardware components can also be considered a form of update. Software or hardware components can be updated for various reasons, including resolving defects, adding new features, or fixing cybersecurity vulnerabilities. ISO/SAE 21434 includes requirements in Clause 13 to address cybersecurity for any updates made after development, whether those updates are to address cybersecurity issues or for other purposes such as resolving defects or adding new features. The intent is to ensure that updates to components, as well as the original components, are developed in accordance with ISO/SAE 21434.

Software updates are the subject of UN Regulation 156, which entered into force at the same time as Regulation 155 for cybersecurity. Regulation 156 contains requirements for vehicle manufacturers to implement a "software update management system" (SUMS) and requirements for the vehicle itself regarding the implementation of software updates. These requirements apply regardless of whether the software is updated by wireless (OTA) or wired methods. At the time of writing, a new international standard, ISO 24089, is under development, which will specify an industry consensus approach to implementing software updates, and can be seen as guidance on meeting the requirements of Regulation 156.

Cybersecurity aspects of software updates, which are part of the scope of UN Regulation 156 and ISO 24089, include protecting the authenticity and integrity of software updates during transmission and installation in the vehicle.

There are also cybersecurity considerations for updates by means of hardware replacement, for example, when an ECU that implements a cryptographic mechanism needs to be replaced, there should be appropriate measures for the provisioning of that ECU in the vehicle, including the initialization of the keys in the new ECU and revocation of the keys in the old ECU.

Besides cybersecurity, there are also additional considerations for updating software, for example, whether the vehicle needs to be stationary for the update to be applied, whether the user of the vehicle needs to be informed or consent to the update, and how to avoid a software update being partially applied, possibly resulting in a nonfunctioning or incorrectly functioning ECU.

End of Cybersecurity Support

Given the requirements above to manage the cybersecurity of vehicles and their components during their operational life, it is clear that manufacturers need to plan how this support will be provided over such long timescales. During the lifetime of a typical vehicle, its manufacturer will introduce many more product lines, which will also need support; supporting a wide range of products, each potentially with multiple possible configurations is, therefore, a major practical challenge.

Recognizing this challenge, ISO/SAE 21434 requires organizations to determine when cybersecurity support for their products will be terminated and to have procedures for notifying customers. Cybersecurity support includes several aspects, including the provision of software updates, monitoring, and incident response. As well as vehicle manufacturers providing this communication to vehicle owners, this communication is also needed throughout the supply chain between suppliers and their customers to ensure that components for which cybersecurity support will end can be managed and appropriate measures taken.

Decommissioning

Decommissioning occurs when a vehicle or component is taken out of service and, unlike the end of cybersecurity support, does not necessarily occur with the manufacturer's knowledge, for example, a faulty ECU that is replaced in a vehicle needs to be decommissioned securely to ensure that discarded components cannot be used to compromise other components. Procedures may therefore need to consider aspects such as how cryptographic keys are revoked or erased from ECUs when they are decommissioned.

Since decommissioning often happens in an uncontrolled environment, it is difficult to enforce the application of such procedures. Therefore, it is generally advisable to consider during development how vehicles and components will be taken out of service and design cybersecurity controls that do not depend on special decommissioning procedures where possible, thus minimizing the risk of cybersecurity being compromised.

The Aftermarket

Another important aspect to consider is aftermarket use cases, which may impact cybersecurity. Section 2.2 has already discussed the implications of "right to repair" legislation; here we are referring to the supply and use of cybersecurity-relevant equipment in the aftermarket.

Vehicle manufacturers (OEMs) already have use cases for aftermarket equipment, which can be relevant for cybersecurity. Some examples include:

1. An OEM develops and markets an aftermarket product even if a supplier is responsible for the detailed product development (e.g., OEM supplied trailer interface module). The OEM (and their supplier) would be responsible for the entire cybersecurity lifecycle.

2. An OEM provides an interface, but an aftermarket installer provides the device (e.g., an OEM-authorized interface to request increased powertrain torque for a power take-off application). The OEM would be responsible for the cybersecurity of the interface, while suppliers and integrators would be responsible for the cybersecurity aspects of the remaining elements. The supporting processes of ISO 26262 Part 8 Clauses 15 and 16 (even though aimed at functional safety and specifically at commercial vehicle applications) may be relevant models for how to interface to/from different applications. This use case may also apply to the "App Store" models referred to above where the "interface" is in effect a software API.
3. There is no OEM involvement (it is purely an aftermarket product and installation). Cybersecurity is entirely the responsibility of the aftermarket actors. However, for an OEM, some caveats about "reasonably foreseeable" use cases might apply. A good example of this is where insurance companies provide "pay as you drive" policies making use of a "dongle" that is installed on the OBD-II port. There are documented cases of security researchers finding significant vulnerabilities in these devices [3.15].

Note that in all these use cases, the ethos of ISO/SAE 21434 would be that the original product developers need to have awareness of the aftermarket use cases and, therefore, its security implications, but they do not execute or control the process or activity themselves. This is similar to the approach to decommissioning (see Section 3.7.8).

Final Thoughts

Volha Hlinskaya/Shutterstock.com.

So we've talked about processes and engineering activities, but how do we know we've done enough and that we've done the right things? In the next chapter, we describe how to establish assurance as a means of demonstrating the achieved cybersecurity in a product.

4
Assurance and Certification

So far, we have explained the need for automotive cybersecurity and how a cybersecurity process gives a managed approach for achieving cybersecurity. But how can we demonstrate that the needs of a specific product have been met? This is the subject of "assurance," In general terms, assurance means justifiable grounds for confidence that the required properties will be achieved subject to a level of risk that they may not be achieved that is acceptable to the stakeholders. The focus of assurance is on activities that provide a sufficient level of confidence, rather than any absolute proof or guarantee.

In the context of cybersecurity, assurance can be considered as a means of establishing confidence that:

- The engineering of the product has taken cybersecurity into account and adequately addresses threats foreseen during development.
- The implementation of the product achieves a level of cybersecurity risk that is acceptable to the stakeholders.
- Appropriate processes are in place for residual risk to monitor and respond to incidents during the operational lifetime and are effective in resolving emerging issues.

The concept of assurance is difficult in the cybersecurity context for several reasons:

- Even at product launch, it is not possible to be confident that all vulnerabilities have been identified and mitigated.

- Vehicles do not operate in isolation but are connected to other systems and a wider environment that effectively widens the opportunities for threat actors.

- During the vehicle's lifetime, threat actors will continue to seek and exploit vulnerabilities, and the technologies used in the product will also change, potentially introducing new vulnerabilities.

Consequently, cybersecurity assurance cannot be achieved by an activity at a single point in the lifecycle of the product; in order to establish and maintain trust, assurance must be built up as an ongoing process and updated at regular intervals throughout the operational life of the vehicle, which can span multiple decades.

Assurance Activities

This section discusses some typical assurance activities and their application to cybersecurity. Generally, these activities are concerned with demonstrating the achievement of cybersecurity and the associated risk reduction and are not simply about confirmation that a product has been completed and implements its requirements but that the product and its requirements achieve the required cybersecurity properties.

The following activities are described and represent the approximate order of completion, although note that some of these activities (such as the assurance case and cybersecurity assessment) are intended to be ongoing rather than simply a "one shot" activity at the end of product development:

- Validation.
- Assurance case.
- Audit.
- Assessment.
- Certification.
- Type approval.

Validation

Ievgeniiya Ocheretna/Shutterstock.com.

"Validation" is a term that is commonly misunderstood. People frequently use the term "verification and validation" as though these two activities are somehow the same or even interchangeable, but in standards such as ISO/SAE 21434 [4.1] and ISO 26262 [4.2], there is a clear distinction made between them:

- Verification is usually concerned with demonstrating that specified requirements have been fulfilled—"did we build the system correctly?".

- Validation is usually concerned with demonstrating that top-level requirements (e.g., the cybersecurity goals) are adequate, have been achieved, and give the required risk reduction—"did we build the correct system?".

Specifically in the context of ISO/SAE 21434, cybersecurity validation is performed on the item in the vehicle in a representative operational environment and production configuration to validate the cybersecurity goals and cybersecurity claims, confirm the item achieves the specified cybersecurity goals, and confirm that no unreasonable risks remain.

Cybersecurity validation builds upon verification activities but will also involve additional activities such as vulnerability testing and penetration testing. In the context of validation, these activities can be performed by independent teams, such as "red teams" or similar groups whose primary purpose is to carry out offensive testing on the organization's products with the goal of identifying previously unknown vulnerabilities.

Independent offensive testing activities can also be useful for identifying new and unforeseen threat scenarios that are either specific to the implementation of the organization's products or in a technology, in general. This adversarial approach to cybersecurity testing is best implemented independently of the development teams to avoid conflicts in objectives and to enable the "red team" to focus on finding issues that need to be resolved, rather than bringing products to market. Large organizations in the technology and other sectors have been operating these teams for several years. High-profile examples include Google Project Zero [4.3] and IBM X-Force [4.4], who are well known for finding classes of vulnerabilities in widely used technologies, protocols, and software, as well as specific vulnerabilities in their product-specific implementations. As such, these teams play a key role in gathering and understanding threat intelligence and can be considered as a source of cybersecurity information for the cybersecurity monitoring activity which was discussed in Section 3.7.1.

Assurance Case

Both ISO 26262 and ISO/SAE 21434 require that a "case" be made for the achieved property (functional safety, cybersecurity) of the product. We have used the term "assurance case" as a more general term to cover the specific instances of a [functional] safety case and a cybersecurity case as these may well share common arguments and evidence (see below).

A general definition of an assurance case is "a structured argument, supported by a body of evidence that provides a compelling, comprehensible and valid case that a system is free from unreasonable risk for a given application in a given operating environment" (adapted from UK Def Stan 00-56 [4.5]).

Another definition can be seen in the goal structuring notation (GSN) community standard [4.6] "A reasoned and compelling argument, supported by a body of evidence, that a system, service or organization will operate as intended for a defined application in a defined environment."

The following significant points should be noted from these definitions:

1. An assurance case is based on evidence, so it is important to be able to identify the activities that have been conducted, and there must be a defined and traceable trail that leads to this evidence.
2. The evidence alone is insufficient; often evidence alone points to an implicit argument for compliance with a standard or process rather than demonstrating why that evidence is proof of the correct activities having been conducted. Therefore, the role of an argument is central, not just in claiming that the activities have been done but also that the right activities have been done in a way that contributes to the overall assurance of the product.
3. An assurance case is based on a defined application and operating environment. This aligns closely with the concept of an item definition used in ISO/SAE 21434 and reflects that the context of an application is of key importance.
4. While these definitions refer to "in a context," the emergent behaviors when the item is taken out of the assumed or defined context may also be important for cybersecurity.
5. Assurance cases are based on achieving freedom from unreasonable risk; such a claim is often placed at the top of an argument structure (see below).
6. Assurance cases need to provide a compelling and comprehensible argument. This implies that the argument is likely to be provided to, and needs to be understood by, one or more stakeholders. Such stakeholders could be internal (such as a product release authority) or external (such as a regulator). Argument-based cases are also extremely useful for those conducting assessments (whether on a first-party, second-party, or third-party basis).

A general model of assurance cases is seen in Figure 4.1. In this model, the claim is the top-level assertion concerning the property of the system, e.g., "the system is

FIGURE 4.1 General model of an assurance case.

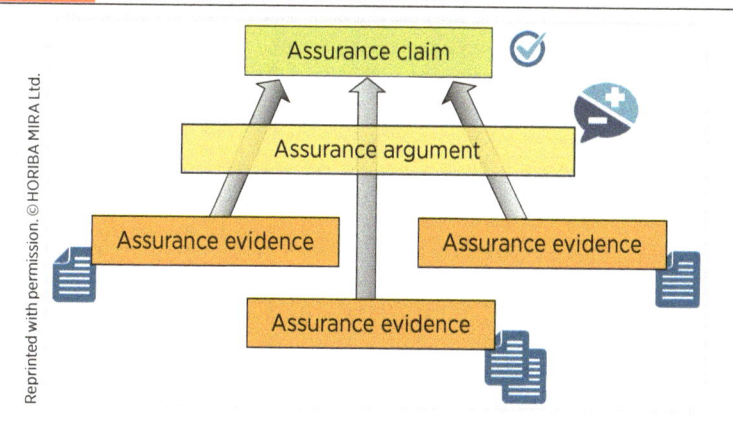

FIGURE 4.2 MISRA layered model for an assurance case.

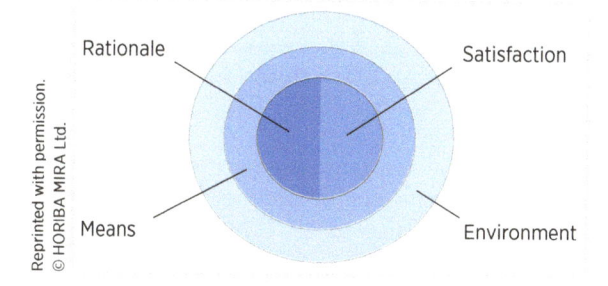

free from unreasonable cybersecurity risks." The evidence is, for example, the work products and other activities required by ISO/SAE 21434, and the argument is concerned with demonstrating how the evidence supports the claim that is being made—not just that the evidence exists.

Standards such as ISO/SAE 21434 and ISO 26262 require an argument-based assurance case, but do not specify how to construct such an argument. Other sources of guidance are available such as the MISRA Guidelines for automotive safety arguments [4.7].

A key aspect of the MISRA guidelines is the use of a layered argument model as shown in Figure 4.2.

The core of this argument model is the "Satisfaction" and "Rationale" themes, and these typically (but not exclusively) are related to the technical design and implementation of the product:

- The "Satisfaction" theme is concerned with claiming that the behavior of a vehicle, system, or element satisfies the cybersecurity requirements allocated to them.

- The "Rationale" theme covers the technical rationale for the adequacy of cybersecurity requirements, risk classifications, CAL ratings, etc. The associated evidence typically points toward the results of an analysis or logical reasoning activity. It is quite often "what engineers have in their heads."

The outer layers of the argument model are the "Means" and "Environment" themes, and these are typically (but not exclusively) related to the process and organizational context in which the activities take place:

- The "Means" theme is concerned with claiming that adequate means have been used to perform a specific activity, for example, developing work products or performing a review. It can be helpful to consider means-based claims relating to "people," "processes," or "tools." Note that the "Means" theme is closely related to ISO/SAE 21434 Clause 6 "Project dependent cybersecurity management."

- The "Environment" theme is concerned with claims regarding the development environment in which activities took place. These claims are not specific to any one particular activity; therefore, these claims are amenable to reuse across arguments for different items. This theme may include claims regarding aspects of whole departments or organizations. Note that the "Environment" theme is closely related to ISO/SAE 21434 Clause 5 "Organizational cybersecurity management."

It is important to note that assurance arguments should consider the role of counterarguments and counterevidence. One of the criticisms frequently levelled at the use of assurance cases is "confirmation bias." This is where engineers assume they have "done the right thing" and, therefore, construct the assurance case accordingly. However, and particularly where the assurance case is to be examined by an independent party such as an assessor or regulator, it is important that the assurance case argues not only why the evidence demonstrates adequate mitigation of risk but also counters claims of inadequate mitigation of risk.

There are many different means of presenting an assurance argument; one technique that is used to provide examples in the MISRA document is GSN [4.6] although it is not necessary to use GSN to construct an argument.

An example of how an assurance argument for a cybersecurity case might be constructed using GSN is shown in Figure 4.3. In this example, the argument is structured around risk treatment decisions for threat scenarios, with the argument for each type of risk treatment developed separately for each of the four subgoals. The "TARA ACP" is an "assurance claim point," as defined in the MISRA Guidelines for automotive safety arguments [4.7], which in this case is intended to link to another part of the argument structure related to the threat analysis and risk assessment. Constructing the GSN diagrams in this way allows the argument to be presented in a clear and comprehensible way without excessive clutter.

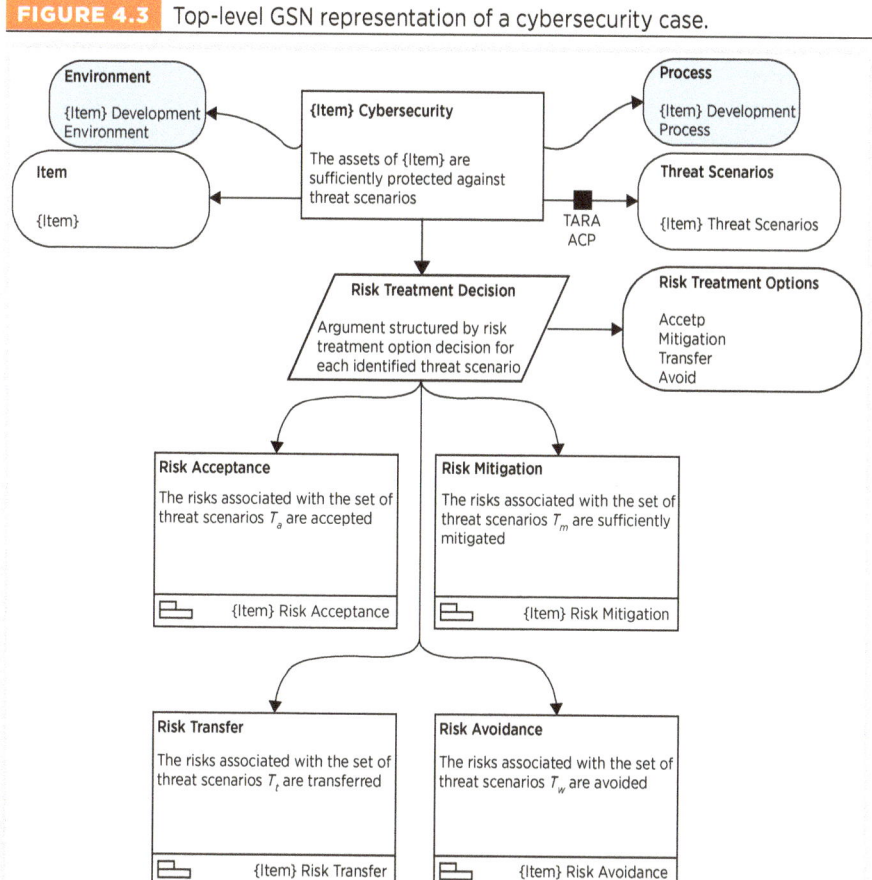

FIGURE 4.3 Top-level GSN representation of a cybersecurity case.

```
Environment                    {Item} Cybersecurity              Process
{Item} Development                                                {Item} Development
Environment                    The assets of {Item} are          Process
                               sufficiently protected against
Item                           threat scenarios                  Threat Scenarios
{Item}                                              TARA         {Item} Threat Scenarios
                                                    ACP

                               Risk Treatment Decision           Risk Treatment Options
                               Argument structured by risk       Accetp
                               treatment option decision for     Mitigation
                               each identified threat scenario   Transfer
                                                                 Avoid

         Risk Acceptance                    Risk Mitigation
         The risks associated with the set of   The risks associated with the set of
         threat scenarios T_a are accepted      threat scenarios T_m are sufficiently
                                                mitigated
              {Item} Risk Acceptance                {Item} Risk Mitigation

         Risk Transfer                        Risk Avoidance
         The risks associated with the set of   The risks associated with the set of
         threat scenarios T_t are transferred   threat scenarios T_w are avoided
              {Item} Risk Transfer                  {Item} Risk Avoidance
```

Reprinted with permission. © HORIBA MIRA Ltd.

Audit

Alfa Photo/Shutterstock.com.

An audit is variously defined as:

- "[a] process for obtaining relevant information about an object of conformity assessment and evaluating it objectively to determine the extent to which specified requirements are fulfilled" (ISO/IEC 17000:2020 [4.8]).
- "examination of an implemented process with regard to the process objectives" (ISO 26262:2018 [4.2]).
- "examination of a process to determine the extent to which the process objectives are achieved" (ISO/SAE 21434:2021 [4.1]).

In practical terms, an audit is usually performed to establish that a process is conformant with a set of requirements (typically those given in a standard). The following general principles should be noted.

An audit may be conducted as a first-party, second-party, or third-party audit. A first-party audit is conducted internally to an organization (albeit usually by an independent internal department such as a quality department) and is usually motivated by being able to demonstrate and improve the organizational capability to meet a set of requirements. A second-party audit is conducted externally to an organization but usually by a party who has an interest in the organization's activities, therefore typically by a customer. A third-party audit is conducted by an independent organization that does not have an interest in the audited organization.

Third-party audits are commonly found in formal accreditation of organizations (often called "certification"), for example, accreditation of an organization against a quality standard such as ISO 9001. Such third-party audits often take the form of a "management system audit." See also Section 4.1.5.

In ISO/SAE 21434, an organizational cybersecurity audit is required. This can be understood as a CSMS audit concerned with establishing an organizational capability to meet the process objectives of the standard. Both ISO/SAE 21434 and ISO 26262 permit an "objectives-oriented" approach to audit rather than simply determining if the processes meet the requirements of the standard, although demonstrating that all requirements are met is one route to demonstrating that the objectives are also met. This "objectives-oriented" approach permits flexibility to adapt to specific organizational practices as well as novel approaches and technologies.

In contrast the audit requirements in ISO 26262 are concerned with demonstrating that conformant processes have been used to develop a particular product ("item"); thus, they are specified as a "functional safety audit" rather than a "functional safety management system audit." In ISO/SAE 21434, the product (item)-specific evidence of application of the processes is implicit in the cybersecurity assessment (q.v.)

ISO/SAE 21434 requires that the audit is conducted with "independence," but this does not necessarily imply that a third-party audit is required. A formal independence scheme is not defined, but a similar scheme to that used in ISO 26262 could be adopted. However, the ISO 26262 independence requirements (and therefore by implication the ISO/SAE 21434 independence requirements) are biased toward a first-party (internal) audit and are not intended to give a mandate for any third-party audit or formal certification. Second-party audits may take place in the context of a customer/supplier relationship. Nevertheless, the UN Regulation 155 requires a certificate of compliance for the CSMS, which is a third-party audit requirement albeit by a Type Approval Authority.

> As a brief aside on terminology it is noted that when an audit is part of a formal accreditation activity, the term "assessment" is used to refer to it. However, there is a clear distinction in technical engineering standards such as ISO/SAE 21434 and ISO 26262 between audit as an activity that examines a process and assessment as an activity that makes a technical judgment on a product.

Assessment

Wright Studio/Shutterstock.com.

A safety assessment is variously defined as:

- "independent … [advocacy] for the level of confidence in the safety delivered to the end customer" [4.9].
- "the formation of a judgement, separate and independent from any system design, development or operational personnel, that the safety requirements for the system are appropriate and adequate for the planned application and that the system satisfies those safety requirements" [4.10].
- "examination of whether a characteristic of an item or element achieves the ISO 26262 objectives" ([4.2]; also with a requirement to "judge the achieved functional safety of the item, or the contribution to the achievement of functional safety by the developed elements").

In contrast the definition of a cybersecurity assessment in ISO/21434 [4.1] is rather brief: "judgement of cybersecurity."

In practical terms, therefore, an assessment is involved with making an independent judgment that a product or a constituent element of a product achieves its attributes (whether safety or security). As a judgment this implies a deep technical review by one or more experts based on available evidence and associated arguments for the adequacy of those activities. A safety case or cybersecurity case based on an argument can be a powerful basis for such an assessment. An assessment will also typically take account of audit results as part of the evidence—these can either be provided as the results of a separate audit or built into the assessment activities.

As with audits, ISO/SAE 21434 requires independence in the cybersecurity assessment but again does not define an independence scheme. The presumption again is that this is a requirement for a first-party assessment although the second-party application may again take place in the context of a customer/supplier relationship. UN Regulation 155 [4.11] effectively requires a third-party evaluation of certain technical aspects by the Type Approval Authority.

Certification

Certification refers to an activity that formally accredits an organization based on the results of a third-party audit or assessment activity. It should be noted that many automotive-industry standards including ISO/SAE 21434 and ISO 26262 are engineering standards, and it is incorrect to refer to these as "certification" standards. There is no mandate for certification in these standards, and there are no formal certification schemes defined for these standards; therefore, any certification offerings are purely commercial propositions of individual certification bodies. The motivation of following these standards is to use them to help develop a safe and secure product, not to be "certified" against the standard per se.

For conformity with standards such as ISO/SAE 21434 and ISO 26262, it is sufficient to conduct audits and assessments with the defined degree of independence, which need not involve an external organization if the required competence and independence can be achieved by internal organizational means. Nevertheless, formal certification can be seen as attractive in the supply chain so that providers of components, particularly those developed as an "element out of context," can demonstrate that a competent third party has reviewed a product.

If formal certification is desired (either to achieve certification or when making use of a certified product), it should be noted that:

- A certification body needs to be accredited themselves to demonstrate their procedures, competence, and impartiality; this accreditation is typically performed by a government body against international accreditation standards. Appropriate accreditation standards in the context of engineering standards such as ISO/SAE 21434 include ISO/IEC 17021 (accreditation of bodies providing audit and certification of management systems—appropriate for audit activities) and ISO/IEC 17065 (accreditation of bodies certifying products, processes, and services—appropriate for both audit and assessment activities).

- Accreditation of a certification body includes demonstrating that the body is impartial—this is a more onerous requirement than the "independence" schemes of ISO/SAE 21434 or ISO 26262. In particular, the impartiality requirements of accreditation standards such as ISO/IEC 17065 mean that the same organization cannot offer both consulting and certification on the same processes or products.

- National accreditation bodies are signatories to international mutual recognition agreements (see, e.g., [4.12]) such that accredited certificates are recognized globally without further certification being required.

Certification is not a "magic bullet" and users of a certified product still have obligations to use the product according to the restrictions associated with the certificate. It is not simply a case of using a "certified" component, and it automatically follows that the product it is integrated into meets the required properties. Instead, the integrator needs to review the information provided as part of a "certification kit" or "pre-qualification evidence" to ensure that the component meets the requirements that will be allocated to it, any constraints in the way that the component will be used are fulfilled, and any capabilities that need to be provided externally to the component are implemented by the integrator.

As an example, a software component such as an operating system or hypervisor may come with a claim of "pre-qualification" or even "certification" against ISO/SAE 21434 or ISO 26262. Such claims must be understood meaning that the product has been developed as an "element out of context", that the requirements of relevant standards have been followed as far as feasible, and that any CAL or ASIL quoted is seen as a capability to support applications with requirements up to that CAL or ASIL. The user is still responsible for configuring the software component in accordance with the integrator guidance, and determining whether the product-specific requirements are capable of being fulfilled by the off-the-shelf component. It is simply not enough to "drop in" a pre-qualified component and assume all will be well, for example, a pre-qualified component may not contain a specific type of cybersecurity control that is needed for a particular application or may require the integrator to provide an additional cybersecurity control.

Figure 4.4, on page 74, illustrates the workflow to make use of a "pre-qualified" or "certified" component developed as a "security element out of context" and makes clear that extensive integrator activities may still be required.

FIGURE 4.4 Overview of workflow for use of pre-developed components.

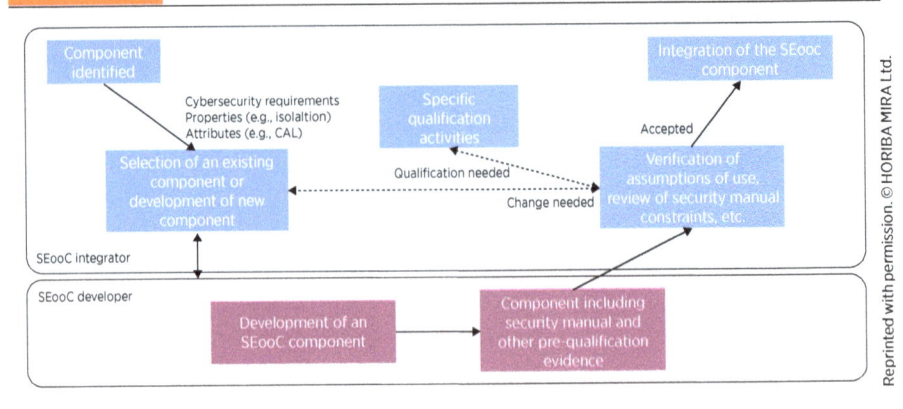

Type Approval

Type Approval is a special case of certification applied to road vehicles in Europe and other territories. It works on the premise of submitting a product representative of the "type" that will go into volume production, along with evidence of conformity of production (namely, that identical copies of the product will be made through the production process). The Type Approval Authority or a designated Technical Service will examine the documentation provided and conduct witnessed tests against the defined requirements of the applicable regulations and then grant approval for the product. Type Approved products may be recognized by application of the "e" or "E" marks, depending on the applicable regime (European or worldwide).

Type Approval requirements are historically seen as setting minimum requirements. A good example of this is seen in the provisions for functional safety of so-called "complex" electronic systems in braking and steering. These regulations do not require the application of ISO 26262 but, instead, represent a high-level set of requirements that the Type Approval Authority will examine at the end of product development along with witnessed testing of safety concepts. However, the newer regulations emerging, including UN Regulation 155 for cybersecurity, will require earlier involvement of the Type Approval Authority and a greater level of technical scrutiny.

The UN Regulations 155 and 156, developed by the UNECE task force CS/OTA (cybersecurity and over the air updates) under WP.29, define new type approval requirements for cybersecurity and OTA updates [4.11]. These regulations entered into force from January 2021 and are being actively adopted by UN contracting parties.

The two regulations are similar in structure, each containing two groups of requirements:

1. A mandatory audit by a Type Approval Authority or technical service of a vehicle manufacturer's CSMS and SUMS, resulting in a cybersecurity certificate of conformity. This must be in place before a vehicle manufacturer can gain Type Approval for a new vehicle.
2. An assessment against the cybersecurity requirements for vehicle type and software update requirements for vehicle types. The assessments are expected to involve a Type Approval Authority verifying that a new vehicle has been appropriately engineered with relevant risks identified, analyzed, and mitigated.

At the time of writing, the UNECE task force CS/OTA is developing a guidelines document, based on the same requirements as the two regulations, which can be used by contracting parties to the UN 1998 agreement. This will mean that the requirements will need to be considered in additional regions, including those that do not follow vehicle Type Approval regimes, such as the USA.

Assurance Summary

Table 4.1 compares and contrasts audit and assessment activities between ISO/SAE 21434 [4.1], UN Regulation 155 [4.11] and ISO 26262 [4.2].

TABLE 4.1 Comparison of assurance approaches in legislation and standards.

Activity or topic	Scope	ISO/SAE 21434 requirements	UN Reg 155 requirements	ISO 26262 requirements
Confirmation reviews	"Mini assessments" of key work products or activities	Not included	Not included	Specific requirements for key deliverables, e.g., safety plan, HARA, functional safety concept
Management system audit	Establish process capability	Organizational cybersecurity audit	CSMS certificate	Not included
Project-specific audit	Establish that conformant processes have been used on a specific product	Implicit in assessment	Implicit in specific product approval (from July 2024)	Functional safety audit
Assessment	Establish that the product meets an adequate level of risk reduction	Cybersecurity assessment	Implicit in specific product approval	Functional safety assessment
Independence	Ensure that the activities have been done without undue bias or pressures from the developers	Requirement stated but no specific details. Informative example scheme related to CAL	Implicit in the Type Approval process	Detailed scheme depending on ASIL of safety goals/safety requirements and the review being conducted
Certification	Third-party audit or assessment by an accredited body	Not required	The Type Approval process	Not required

FIGURE 4.5 Relationship between ISO/SAE 21434 and UN Regulation 155.

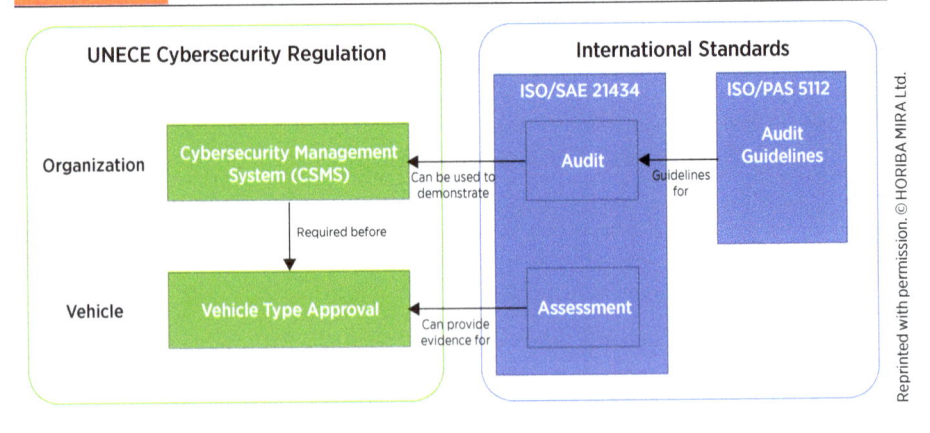

Reprinted with permission. © HORIBA MIRA Ltd.

Figure 4.5 shows how the assurance activities contained in international standards can be used to support regulatory approvals:

Final Thoughts

margotikaj/Shutterstock.com.

So there it is, we've talked about how to establish assurance as a means of demonstrating the achieved cybersecurity in a product. In the next chapter, we'll draw some final conclusions and make recommendations for going further.

5

Conclusions and Going Further

Cybersecurity presents major technical and nontechnical challenges for modern cyber-physical systems, and we have explored the specific manifestations of these challenges for the automotive industry as vehicles become increasingly connected and automated. With cybersecurity being a relatively new discipline for automotive manufacturers, we outlined how it is related to other disciplines, especially functional safety. The need to address potential safety-related consequences of cyber-attacks and the ability of an attacker to deliberately cause faults in electronic systems means that functional safety and cybersecurity should not be considered in isolation from each other. Indeed there are practical opportunities to align the process frameworks and manage the conflicts and synergies that arise from the needs of both areas, while still ensuring the unique requirements of each discipline are met.

We have seen how cybersecurity presents a number of challenges as an engineering discipline, particularly in the automotive domain, including the need to ensure that systems can remain resilient over vehicle lifetimes stretching into decades; how to update software on vehicle systems reliably, safely, and securely; the scalability of cybersecurity controls established in other domains to embedded automotive systems, which are typically resource constrained in terms of performance, memory, and bandwidth; and the need to handle changes of vehicle ownership, end of life, and decommissioning, especially regarding the handling of personal data.

The automotive threat landscape presents further challenges, with the need to address not only the more familiar financial and privacy aspects of cyber-attacks but also threats leading to safety and operational impacts. The vehicle attack surface offers a wide range of potential entry points for attackers to realize these threats, including wireless communication interfaces using common technology such as Wi-Fi and Bluetooth; however, additional attack vectors include more automotive-specific interfaces and environmental sensors, such as radar, lidar, and cameras, which are not yet comprehensively explored or understood in terms of their attack potential. Further attack vectors exist that require wired or physical access to vehicle interfaces or components, such as the ubiquitous OBD-II port, attacks via in-vehicle network wiring, or hardware attacks on ECUs. As we have seen, practical attacks do not always involve a single action exploiting a single vulnerability; typically, an attacker will need to chain together attack vectors, exploiting multiple vulnerabilities that may in isolation be considered benign. These facts mean that cybersecurity does not lend itself to being decomposed into subcomponents to be addressed in isolation but rather needs a holistic approach and an overall cybersecurity concept managed throughout the lifetime of the vehicle.

We have introduced how cybersecurity is not a purely technical discipline but requires a combination of people, process, and technology elements to be brought together under a common framework. Chapter 3 outlined the key elements needed to establish a cybersecurity process, including an overview of the requirements of automotive cybersecurity regulations, standards, and best practices and the different lifecycle models to which such a process could be applied.

We have seen how ISO/SAE 21434 provides a framework for organizations to implement a dual approach, encompassing both proactive and reactive cybersecurity engineering and a set of organizational management activities. The management and governance activities ensure that the process and people aspects of cybersecurity are adequately addressed, including assigning responsibilities and resources at all levels of the organization, ensuring people have the right competencies and that a cybersecurity culture is embedded across the organization.

A proactive "security by design" approach to engineering is critical to ensure that threats are appropriately identified and risks assessed and treated from the beginning of the design process and that appropriate and effective cybersecurity controls are built into the vehicle and its systems, verified, and validated before production starts. We explained the dynamic nature of cybersecurity risk and the role of the risk management concept at the core of ISO/SAE 21434, including how risk assessment must be iteratively applied, for example, as an initial concept phase TARA and as part of ongoing vulnerability management activities throughout development and beyond. The role of verification and validation methods, including analysis and testing, in confirming the correctness and effectiveness of the implemented cybersecurity controls and identifying any previously unknown vulnerabilities was also explored, including how the CALs can be used to determine the level of effort or rigor apply to particularly high-effort cybersecurity activities such as penetration testing.

Cybersecurity does not stop at the end of product development, and we outlined some considerations for the production environment, including implementing any cybersecurity requirements for production such as injection of cryptographic keys and any requirements related to tools, equipment, or their configuration.

Even once the vehicle has been produced and enters the operation phase, we still cannot consider cybersecurity activities are complete. We have seen that due to the dynamic and evolving nature of threats, monitoring activities must continue through which cybersecurity information is collected and processed and its relevance to the organization's products is evaluated. This enables the organization to become aware of emerging threats, vulnerabilities, and attacks that affect its products and react accordingly through incident response activities. These may include developing and deploying updates for the vehicle or off-board systems, initiating change management for products in development and public relations or other communications activities when required.

Beyond monitoring sources of threat intelligence, the UN Regulation 155 further requires that attacks are detected and responded to in an appropriate timeframe. This means that the cybersecurity concept should include controls that not only aim at preventing known attacks but also enable the detection of attacks. Achieving comprehensive real-time detection and response to attacks is challenging and still requires significant research and development; however, approaches involving the off-board monitoring and handling of events and incidents by means of solutions such as a VSOC are likely to provide a practical means to start implementing this kind of cybersecurity operations capability.

In Chapter 4 we introduced the concept of cybersecurity assurance, how it can be achieved, and a review of the various assurance methods appropriate for cybersecurity. These include validation of the cybersecurity goals, using methods such as penetration testing and red teams. We explained the role of an assurance case, which provides a structured argument with supporting evidence for the adequacy of the cybersecurity activities carried out for a vehicle or system.

Audit and assessment are typical assurance activities, and we have explored how they can be used as both internal activities, as well as third-party activities, for example, in a type approval context. The role of certification was also introduced, including the benefits and limitations of pre-qualified or certified products. We also explained the special case of certification, type approval, which is particularly relevant for cybersecurity in countries adopting UN Regulation 155.

In summary, in this introduction to automotive cybersecurity, we aimed to provide an overview of how to start implementing cybersecurity activities within an organization in line with the requirements of ISO/SAE 21434. There is of course much that we have not been able to cover in this short book, and the evolving nature of cybersecurity that we have returned to at multiple points throughout this book means that new requirements, improved techniques, and deeper practical experience will emerge over time.

As practitioners beginning to implement the activities described, a number of practical challenges will need to be resolved, including:

- How to implement a risk-based cybersecurity engineering process with a limited supply of experts.
- How to manage the effort of analysis-heavy processes like risk assessment and vulnerability management.

- How to determine where to allocate the effort and how much is enough—including the use of the CAL concept to determine and implement this.
- How to measure and assess the ongoing effectiveness of cybersecurity controls in the face of an evolving threat landscape.
- How to measure and assess the ongoing effectiveness of operational detection and response capabilities to ensure that we can react to new threats.
- How to move operational capabilities toward real-time detection and response, recognizing that detecting and responding in a centralized way such as a VSOC may be too late for some scenarios.
- How to develop, build, verify, and validate resilient self-healing systems.

Implementing a comprehensive approach to cybersecurity is a major business transformation activity, and we have seen from the wide range of activities required by automotive legislation and standards that this will be a significant challenge for many organizations. Although regulatory compliance will be a strong driver for automotive cybersecurity, meeting the requirements of the regulation should be seen as a minimum starting point, and compliance activities alone will not always be sufficient to ensure that vehicles remain safe and secure against cyber-attacks.

It is often said that cybersecurity is like a chain in that it is only as strong as the weakest link: an assertion backed up by the frequent emergence of newly discovered attacks and vulnerabilities. However, cybersecurity capabilities can be developed progressively and iteratively, starting with activities that make the biggest impact. It is always better to put some aspects of a cybersecurity process in place, while others take longer to develop, rather than wait until all aspects are ready to deploy at once. The rapidly developing nature of connected and automated vehicles together with the hostile and dynamic threat environment in which they operate presents a formidable challenge, but one that can and must be addressed through the transformation of automotive engineering practices. This will not be achieved through a single event or step change, but it is a journey—and it is not too late to start.

Frequently Asked Questions

sdecoret/Shutterstock.com.

We conclude with a number of frequently asked questions (FAQs) concerning automotive cybersecurity, ISO/SAE 21434, and wider topics.

What Is the Difference between UN Regulation 155 and ISO/SAE 21434?

UN Regulation 155 is an internationally-adopted regulation specifying cybersecurity requirements for whole vehicle-type approval in those countries that are signatories to the regulation, whereas ISO/SAE 21434 is an international standard specifying industry-agreed best practice and a "common language." UN Regulation 155 and ISO/SAE 21434 do not refer to each other specifically; however, ISO/SAE 21434 can be used to implement a cybersecurity management system as required by the regulation and to demonstrate compliance with the regulation's requirements for vehicle types. Furthermore, there is an "interpretation document" associated with Regulation 155 that may be used by type approval authorities in applying the Regulation that does refer to ISO/SAE 21434.

To Which Types of Vehicles Does UN Regulation 155 Apply?

The regulation applies to passenger cars, buses, and goods-carrying vehicles. It also applies to trailers fitted with at least one ECU and to certain types of light four-wheeled vehicles, if they are equipped with Level 3 or above automated driving functions. Since the regulation applies to vehicles, it directly affects vehicle manufacturers; component suppliers are not directly bound by the requirements of the regulation, although many of the requirements can be considered indirectly applicable since certain activities that enable the vehicle manufacturer to achieve the requirements will be "flowed down" from vehicle manufacturers to their suppliers.

To Which Types of Organization Does ISO/SAE 21434 Apply?

ISO/SAE 21434 applies to the electronic and electrical systems of series production road vehicles and, as such, applies to vehicle manufacturers, as well as the entire supply chain responsible for manufacturing electronic and electrical systems and their components. This includes sub-system and ECU manufacturers, software suppliers, semiconductor manufacturers, and others. The requirements of ISO/SAE 21434 can be tailored according to the type of organization applying the standard since different parts of the standard are applicable to different organizations and their products.

How Do You Audit for Conformance to ISO/SAE 21434?

As we explain in Chapter 4, ISO/SAE 21434 requires an audit of the organization's cybersecurity processes, which can also be understood as a CSMS audit, as also required by UN Regulation 155. While the UN Regulation requires a third-party audit by a Type Approval Authority, ISO/SAE 21434 also accommodates first-party (internal) or second-party (customer/supplier) audits. A further document, ISO/PAS 5112, is under development at time of writing, which will provide further guidance on conducting all these types of cybersecurity audits.

Is It Mandatory to Be Certified against ISO/SAE 21434?

In Chapter 4 we explain that ISO/SAE 21434 is an engineering standard and no formal certification scheme is defined, with any advertised certification offerings being simply optional commercial services. ISO/SAE 21434 permits audits to be carried out without formal certification and internal to the organization, provided the independence requirements can be met. However, sometimes certification can be considered attractive or beneficial for component suppliers to demonstrate that their product has been reviewed by a competent and impartial third party.

Do I Have to Use ISO/SAE 21434 for My Cybersecurity Processes?

As noted above, ISO/SAE 21434 is not currently mandated by any legislation although this may change in the future. An organization may choose to use different processes or approaches, but ISO/SAE 21434 will rapidly become established as the "state of the art." From a practical perspective, if an organization has existing, effective, processes, these could be incorporated as part of an "objectives-oriented" approach to conformance (e.g., in an audit; see Chapter 4). However, ISO/SAE 21434 will become the accepted industry approach in the same way that ISO 26262 has for functional safety, so following the standard adopters will be in step with accepted industry approaches and practices.

How Do I Know If My Item or Component Is Cybersecurity Relevant?

Determining whether an item or component is cybersecurity relevant is one of the first activities to be performed in a development project, as we see in Chapter 3. ISO/SAE 21434 provides some guidance on how to assess cybersecurity relevance in Annex D in the form of a flowchart, which includes criteria such as whether the item or component is safety related or is connected via a network to other components. Organizations can also use the experience of previous products to determine relevance, and although no guidance on this approach is provided in ISO/SAE 21434, it is conceivable that organizations could develop policies for which of their product types are cybersecurity relevant.

The Various Analysis Activities for Cybersecurity Engineering Look Very Time Consuming; How Do I Know When I Have Done Enough?

This is a key question facing the industry in terms of the practical application of the engineering process requirements of ISO/SAE 21434. The activities set out in the standard each contributes to the overall assurance that can be gained in the cybersecurity of the vehicle or component; however, there is no normative requirement on the extent, depth, or rigor of many of the activities, such as attack path analysis or verification activities. The informative concept of CALs is provided in the standard as a means of scaling the effort required for cybersecurity engineering activities based on the level of assurance required. The premise is that the most critical assets need greater assurance and should be subject to a higher level of process rigor to provide the confidence that the assets are sufficiently protected against the relevant threats. The CAL is thus determined based on the threat scenarios and used as an index to select appropriate methods and parameters for performing the corresponding activities to specify, implement, and verify the cybersecurity controls.

Does ISO/SAE 21434 Define Which Cybersecurity Tests Should Be Carried Out?

ISO/SAE 21434 and UN Regulation 155 both require testing to be carried out as part of verifying the effectiveness of the implemented cybersecurity controls. While ISO/SAE 21434 provides some guidance on the use of testing as part of cybersecurity verification and validation, there are no prescriptive requirements for specific tests to be carried out. As we explain in Chapter 3, due to the diverse nature of automotive technology and the evolving threat landscape, cybersecurity tests should not be standardized in a fixed and prescriptive manner since this would not fully enable the effectiveness of the cybersecurity controls to be verified or enable previously unknown vulnerabilities to be identified. Instead, a cybersecurity verification plan should be developed specifically for the item or component, informed by the threat analysis and risk assessment, and specifying an appropriate combination of analysis and test methods for the cybersecurity controls and the threat scenarios they are intended to mitigate.

Final Thoughts

GoodStudio/Shutterstock.com.

We hope you've found this introduction to automotive cybersecurity useful and it will help you in charting a course through legislation and standards ... and, ultimately, in developing and maintaining safe and secure products!

References

Chapter 1

1.1. Federal Register, "Exemption to Prohibition on Circumvention of Copyright Protection Systems for Access Control Technologies," accessed August 26, 2021 https://www.federalregister.gov/d/2015-27212/p-173.

1.2. Checkoway, S., McCoy, D., Kantor, B., Anderson, D. et al., "Comprehensive Experimental Analyses of Automotive Attack Surfaces," in *USENIX Security Symposium*, San Francisco, CA, August 2011.

1.3. Kosher, K., Czeskis, A., Roesner, F., Patel, S. et al., "Experimental Security Analysis of a Modern Automobile," in *The IEEE Symposium on Security and Privacy*, Berkeley/Oakland, CA, May 16-19, 2010.

1.4. Garcia, F.D., Oswald, D., Kasper, T. and Pavlidès, P., "Lock It and Still Lose It-on the (In) Security of Automotive Remote Keyless Entry Systems," in *USENIX Security Symposium*, Austin, TX, August 2016.

1.5. ISO/SAE, "Road Vehicles—Cybersecurity Engineering," ISO/SAE 21434:2021, Revised 2021.

Chapter 2

2.1. BBC, "BMW Fixes Security Flaw That Left Locks Open to Hackers," accessed August 2021, https://www.bbc.co.uk/news/technology-31093065.

2.2. OpenBTS, "Open Source Cellular Infrastructure," accessed August 2021, http://openbts.org/.

2.3. 3GPP, "V2X," accessed August 2021, https://www.3gpp.org/v2x.

2.4. ETSI, "Technical Committee (TC) Intelligent Transport Systems (ITS)," accessed August 2021, https://www.etsi.org/committee/1402-its.

2.5. IEEE, "IEEE 1609.2-2016—IEEE Standard for Wireless Access in Vehicular Environments—Security Services for Applications and Management Messages," accessed August 2021, https://standards.ieee.o20rg/standard/1609_2-2016.html.

2.6. Fernandes, E., Crispo, B., and Conti, M., "FM 99.9, Radio Virus: Exploiting FM Radio Broadcasts for Malware Deployment," *IEEE Transactions on Information Forensics and Security* 8, no. 6 (2013): 1027-1037, doi:10.1109/TIFS.2013.2259818.

2.7. Cheah, M., Shaikh, S., Bryans, J., and Wooderson, P., "Building an Automotive Security Assurance Case Using Systematic Security Evaluations," *Computers and Security* 77 (2018): 360-379, doi:10.1016/j.cose.2018.04.008.

2.8. Miller, C., and Valasek, C., "Remote Exploitation of an Unaltered Passenger Vehicle," in *Black Hat USA*, Las Vegas, NV, 2015.

2.9. Vanhoef, M. and Piessens, F., "Key Reinstallation Attacks: Forcing Nonce Reuse in WPA2," in *ACM Proceedings of the 24th ACM Conference on Computer and Communications Security (CCS)*, Dallas, TX, 2017.

2.10. Rouf, I. et al., "Security and Privacy Vulnerabilities of In-Car Wireless Networks: A Tire Pressure Monitoring System Case Study," in *Proceedings of the 19th USENIX Conference on Security*, Washington, DC, August 2010.

2.11. Garcia, F., Oswald, D., Kasper, T., and Pavlidès, P., "Lock It and Still Lose It—On the (In)Security of Automotive Remote Keyless Entry Systems," in *USENIX Security Symposium*, Vancouver, BC, Canada, 2017.

2.12. Francillon, A., Danev, B., and Capkun, S., "Relay Attacks on Passive Keyless Entry and Start Systems in Modern Cars," *IACR Cryptol*, ePrint Arch. (2011): 332.

2.13. Xu, W., Yan, C., Jia, W., Ji, X. et al., "Analyzing and Enhancing the Security of Ultrasonic Sensors for Autonomous Vehicles," *IEEE Internet of Things Journal* 5, no. 6 (2018): 5015-5029.

2.14. Komissarov, R. and Wool, A., "Spoofing Attacks against Vehicular FMCW Radar," ArXiv, abs/2104.13318, 2021.

2.15. Petit, J., Stottelaar, B., and Feiri, M., "Remote Attacks on Automated Vehicles Sensors : Experiments on Camera and LiDAR," in *Black Hat USA*, Las Vegas, NV, 2015.

2.16. Acharya, S., Dvorkin, Y., Pandžić, H., and Karri, R., "Cybersecurity of Smart Electric Vehicle Charging: A Power Grid Perspective," *IEEE Access* 8 (2020): 214434-214453, doi:10.1109/ACCESS.2020.3041074.

2.17. ISO, "Road Vehicles—Vehicle-to-Grid Communication Interface—Part 2: Network and Application Protocol Requirements," ISO 15118-2:2014, Revised 2014.

2.18. SAE International, "SAE International to Launch Industry-Driven SAE EV Charging Public Key Infrastructure Project," accessed August 26, 2021, https://www.sae.org/news/pressroom/2020/05/sae-international-to-launch-industry-driven-sae-ev-charging-public-key-infrastructure-project.

2.19. Checkoway, S., McCoy, D., Kantor, B., Anderson, D. et al., "Comprehensive Experimental Analyses of Automotive Attack Surfaces," in *SEC'11 Proceedings of the 20th USENIX Conference on Security*, Berkeley, CA, 2011.

2.20. SRLabs, "Wiki—BadUSB Exposure," accessed August 26, 2021, https://opensource.srlabs.de/projects/badusb.

2.21. SAE International Surface Vehicle Recommended Practice, "Diagnostic Connector," SAE Standard J1962, Revised July 2016.

2.22. Miller, C. and Valasek, C., "Adventures in Automotive Networks and Control Units," in *DEFCON 21*, Las Vegas, NV, 2013.

2.23. Kosher, K., Czeskis, A., Roesner, F., Patel, S. et al., "Experimental Security Analysis of a Modern Automobile," in *The IEEE Symposium on Security and Privacy*, Berkeley/Oakland, CA, May 16-19, 2010.

2.24. ISO, "Road Vehicles—Unified Diagnostic Services (UDS)—Part 1: Application Layer," ISO 14229-1:2020, Revised 2020.

2.25. Murvay, P. and Groza, B., "Practical Security Exploits of the FlexRay In-Vehicle Communication Protocol," in *CRiSIS*, 2018, doi:10.1007/978-3-030-12143-3_15.

2.26. IEEE802.org, "P802.1DG—TSN Profile for Automotive In-Vehicle Ethernet Communications," accessed August 26, 2021, https://1.ieee802.org.

2.27. Jagielski, M., Oprea, A., Biggio, B., Liu, C. et al., "Manipulating Machine Learning: Poisoning Attacks and Countermeasures for Regression Learning," in *2018 IEEE Symposium on Security and Privacy*, San Francisco, CA, 2018, 19-35, arXiv:1804.00308, doi:10.1109/sp.2018.0005.

2.28. Evtimov, I., Eykholt, K., Fernandes, E., Kohno, T. et al., "Robust Physical-World Attacks on Deep Learning Models," in *IEEE/CVF Conference on Computer Vision and Pattern Recognition (CVPR)*, Seattle, WA, 2017.

2.29. Herley, C. and Van Oorschot, P., "SoK: Science, Security and the Elusive Goal of Security as a Scientific Pursuit," in *2017 IEEE Symposium on Security and Privacy (SP)*, San Jose, CA, 2017, 99-120, doi:10.1109/SP.2017.38.

2.30. AUTOSAR, "Specification of Secure Onboard Communication," Version 4.3.1, accessed August 27, 2021, https://www.autosar.org/fileadmin/user_upload/standards/classic/4-3/AUTOSAR_SWS_SecureOnboardCommunication.pdf.

2.31. AUTOSAR, "Specification of Secure Hardware Extensions," Version 4.3.1, accessed August 27, 2021, https://www.autosar.org/fileadmin/user_upload/standards/foundation/19-11/AUTOSAR_TR_SecureHardwareExtensions.pdf.

2.32. Henniger, O., Ruddle, A., Seudié, H., Weyl, B. et al., "Securing Vehicular On-Board IT Systems: The EVITA Project," in *25th VDI/VW-Gemeinschaftstagung Automotive Security*, Ingolstadt, Germany, 2009.

2.33. SAE International Surface Vehicle Recommended Practice, "Hardware Protected Security for Ground Vehicles," SAE Standard J3101, Revised February 2020.

2.34. Schneier on Security, "Machine Learning to Detect Software Vulnerabilities," accessed August 26, 2021, https://www.schneier.com/blog/archives/2019/01/machine_learnin.html.

2.35. Hafeez, A., Rehman, K., and Malik, H., "State of the Art Survey on Comparison of Physical Fingerprinting-Based Intrusion Detection Techniques for In-Vehicle Security," SAE Technical Paper 2020-01-0721, 2020, https://doi.org/10.4271/2020-01-0721.

Chapter 3

3.1. MISRA, *MISRA C:2012 Guidelines for the Use of the C Language in Critical Systems*, 3rd ed., First Revision, (Nuneaton: HORIBA MIRA Limited, 2019), ISBN:978-1-906400-21-7.

3.2. Auto-ISAC, "Automotive Information Sharing and Analysis Centre," accessed August 26, 2021, https://www.automotiveisac.com/.

3.3. NHTSA, "Cybersecurity Best Practices for the Safety of Modern Vehicles," accessed August 26, 2021, https://www.nhtsa.gov/sites/nhtsa.gov/files/documents/vehicle_cybersecurity_best_practices_01072021.pdf.

References

3.4. Shostack, A., *Threat Modelling: Designing for Security* (Indianapolis, IN: Wiley, 2014)

3.5. Schneier, B., *Secrets and Lies—Digital Security in a Networked World* (New York: Wiley, 2000), Chapter 21.

3.6. Forum of Incident Response and Security Teams (FIRST), "Common Vulnerability Scoring System (CVSS)," accessed August 26, 2021, https://www.first.org/cvss/.

3.7. Ruddle, A. et al., "Security Requirements for Automotive On-Board Networks Based on Dark-Side Scenarios (EVITA Deliverable 2.3)," EVITA EC Project, V. 1.1, December 30, 2009, doi:10.5281/zenodo.1188418.

3.8. SAE International Surface Vehicle Recommended Practice, "Cybersecurity Guidebook for Cyber-Physical Vehicle Systems," SAE Standard J3061, Revised January 2016.

3.9. ISO/SAE, "Road Vehicles—Cybersecurity Engineering," ISO/SAE 21434:2021, Revised 2021.

3.10. "Common Criteria for Information Technology Security Evaluation," 3 Parts, v. 3.1, Release 5, accessed August 26, 2021, https://www.commoncriteriaportal.org/cc/.

3.11. Wooderson, P. and Ward, D., "Cybersecurity Testing and Validation," SAE Technical Paper 2017-01-1655, 2017, https://doi.org/10.4271/2017-01-1655.

3.12. NIST, "National Vulnerability Database," accessed August 26, 2021, https://nvd.nist.gov/.

3.13. MITRE Corporation, "Common Vulnerabilities and Exposure," accessed August 26, 2021, https://cve.mitre.org/.

3.14. NCSC, "Effective Steps to Cyber Exercise Creation," accessed August 26, 2021, https://www.ncsc.gov.uk/guidance/effective-steps-to-cyber-exercise-creation/.

3.15. Argus Cyber Security, "A Remote Attack on an Aftermarket Telematics Service," accessed August 26, 2021, https://argus-sec.com/remote-attack-aftermarket-telematics-service/.

Chapter 4

4.1. ISO/SAE, "Road Vehicles—Cybersecurity Engineering," ISO/SAE 21434:2021, Revised 2021.

4.2. ISO, "Road Vehicles—Functional Safety," ISO 26262:2018, Revised 2018.

4.3. Google, "Project Zero," accessed August 26, 2021, https://googleprojectzero.blogspot.com/.

4.4. IBM, "IBM X-Force," accessed August 26, 2021, https://www.ibm.com/security/xforce.

4.5. UK Ministry of Defence, "Safety Management Requirements for Defence Systems: Part 1: Requirements," Defence Standard 00-56 Part 1, February 2017.

4.6. Safety Critical Systems Club Assurance Case Working Group, "Goal Structuring Notation Community Standard," Version 3, May 2021.

4.7. MISRA, *Guidelines for Automotive Safety Arguments* (Nuneaton: HORIBA MIRA Limited, 2019), ISBN:978-1-906400-23-1.

4.8. ISO/IEC, "Conformity Assessment—Vocabulary and General Principles," ISO/IEC 17000:2020, Revised 2020.

4.9. MISRA, *Development Guidelines for Vehicle Based Software* (Nuneaton: The Motor Industry Research Association, 1994), ISBN:978-0-9524156-0-2.

4.10. The Institution of Engineering and Technology, "What Is Independent Safety Assurance (ISA)?," accessed September 30, 2021, https://www.theiet.org/impact-society/factfiles/isa-factfiles/what-is-isa/.

4.11. United Nations Economic Commission for Europe, "Addenda to the 1958 Agreement (Regulations 141-160)," accessed September 30, 2021, https://unece.org/transport/vehicle-regulations-wp29/standards/addenda-1958-agreement-regulations-141-160.

4.12. International Accreditation Forum, "About IAF," accessed September 30, 2021, https://iaf.nu/en/home/.

Index

A
"Agile," 36
Arbitrary code, 2
Artificial intelligence, 3, 28, 33
Assets, 2, 5
  developers of, 1
Assurance activities
  assessment, 71–72
  assurance case
    general model of, 66–67
    MISRA layered model, 67
    pre-developed components, workflow for use of, 73, 74
    top-level GSN representation, 68, 69
  audit, 69–71
  certification, 72–74
  concept of, 63–64
  ISO/SAE 21434 and UN Regulation 155, 76
  in legislation and standards, 75
  Type Approval, 74–75
  validation, 64–65
Attack trees method, 47
Automotive cyber-physical systems
  "App Store" models, 16
  attack paths and stepping stones, 27–28
  challenges, 15
  circumvention, 17

cybersecurity engineering, 30–31
end of life, vehicle, 17
and functional safety relationship attributes, 8
electronic steering column lock (ESCL), 9, 12, 13
freedom from interference, 14
hazard analysis and risk assessment (H&R/HARA), 8, 9, 11
industry-agreed method, of risk prioritization, 15
ISO 26262 approach, 11
ISO 26262 Edition 2, 10
isolation, 14
mechanical failure mode, 9
safety goals, 9
safety of the intended functionality (SOTIF), 10
scope restrictions, 9
software architecture, independence in, 14
steering column, 9–10
Threat Analysis and Risk Assessment (TARA), 11
time-sensitive fault detection mechanism, 13

traditional vehicle engineering, 8
legislation, 16–17
management, of cybersecurity, 29–30
requirement skills, for cybersecurity, 31–32
scalability, 15
"software-defined car," 16
software updates, 16
technology, 32–33
vehicle attack surface, 7
  ECUs, 25–27
  in-vehicle networks, 24–25
  wireless interfaces, 18–22
vehicle lifetimes, 15
vehicle ownership, changing models of, 16
Automotive Ethernet, 25
Automotive Safety Integrity Level (ASIL) values, 8, 11–13
AUTOSAR Secure On-board Communications specification, 32

B
"Bare metal" software, 26
Bluetooth, 20

C
Common Vulnerability Scoring System (CVSS), 47
Communication facilities, 3

©2022 SAE International

"Conformation bias," 68
Connected car services, 18
Controller Area Network (CAN), 23–25
"Cybernetics," 2
Cyber-physical systems
  automotive cyber-physical systems (see Automotive cyber-physical systems)
  security engineering for, 2
Cybersecurity, 2
  analysis activities for, 82–83
  assets, developers of, 1
  automotive industry, 3–4
  automotive threat landscape, 77
  CALs, 83
  concept phase
    CAL, 48–49
    design verification, 51
    item definition, 46
    requirements and controls, 49–51
    risk treatment and cybersecurity goals, 47–48
    threat analysis and risk assessment, 46–47
  concepts, 4–6
  cyber-physical systems, 4
  definitions, 4–6
  general aspects of, 35–36
  ISO/SAE 21434, 35, 78, 81–82
  legislation, standards and guidelines, 37
  lifecycle, 37–40
  management
    awareness and competence, 41
    continuous improvement, 42
    culture, 40
    information sharing, 42
    roles and responsibilities, 41
    top management commitment, 40
    penetration testing and red teams, 79
  proactive cybersecurity engineering
    planning, 44–45
    responsibilities at project level, 43–44
  proactive "security by design" approach, 78
  reactive cybersecurity engineering
    aftermarket, 61–62
    attacks, detecting and responding to, 58
    decommissioning, 61
    detection and response processes, 59–60
    event evaluation, 57–58
    incident response, 58–59
    monitoring process, 56–57
    support, end of, 61
    updates, 60
  relevance in, 82
  risk, 6
  security-related incidents, 1
  security researchers, 1
  standards and best practice, 36–37
  technical and nontechnical challenges, 77
  testing
    challenges, 51–52
    correctness testing, 53
    development phases, 53
    functional testing, 53
    penetration testing, 54
    pre-testing analysis, 53
    during production, 55–56
    vulnerability analysis and management, 54–55
  tests, 83
  threat scenarios, 5
  UN Regulation 155, 81
"Cybersecurity assurance level" (CAL), 11–13, 48–49, 83
Cybersecdurity management system (CSMS), 55
"Cyberspace," 2

D
Denial-of-service attacks, 1
Digital Audio Broadcasting (DAB) radio, 20
Digital Millennium Copyright Act (DMCA), 17
Digital security, 2

E
Electronic steering column lock (ESCL), 9, 12, 13
"Environment" theme, 68
Evaluation Assurance Levels (EAL), 49

F
Fault injection attacks, 27
FlexRay protocol, 24
Frequency Modulation (FM), 20

G
Global Navigation Satellite System (GNSS), 19
Goal structuring notation (GSN) community standard, 66, 68
Google Project Zero, 65

H
"Hands-of" regulatory approach, 16
"Harm," 9
Hazard analysis and risk assessment (H&R/HARA), 8, 9, 11
"Heartbleed" SSL vulnerability, 2

Index

I
IBM X-Force, 65
Independent offensive testing activities, 65
Information security, 4
Information Sharing and Analysis Centers (ISACs), 30
Infotainment system, 26
Intangible assets, 5
ISO standards, 12

K
"KRACK" Wi-Fi key reinstallation attacks, 2
"Kubernetes," 2

L
Local Interconnect Network (LIN), 24, 25

M
Malicious data insertion, 2
"Means" theme, 68
Microcontrollers, 3

O
On-Board Diagnostics (OBD)-II port, 23
Original equipment manufacturer (OEM), 3, 61–62
"Over the air" (OTA) software, 16, 29

P
"Power by the hour" scheme, 16

R
"Rationale" theme, 68
Remote keyless entry systems, 21

S
"Satisfaction" theme, 67
Security, 1. *See also* Cybersecurity
  risk, 4
  traditional approaches to, 4
Security operations centers (SOCs), 58
Side-channel attacks, 26–27
Software defects exoploitation, 2
Software-defined radio (SDR), 19
"Software update management system" (SUMS), 60
"Spectre" and "meltdown" vulnerabilities, 2
STRIDE threat modelling method, 47
System-on-chip (SoC)-based systems, 25, 26

T
"Tailoring," 36
Tangible assets, 5
"TARA ACP" assurance claim point, 68
Threat Analysis and Risk Assessment (TARA), 11
Tire pressure monitoring systems (TPMS), 21

V
Vehicles
  attack surface, 7
  ECUs, 25–27
  in-vehicle networks, 24–25
  wireless interfaces, 18–22
  end of life, 17
  lifetimes, 15
  manufacturers, aftermarket use cases, 61–62
  motion functionality, 3
"Vehicle security operations center" (VSOC), 58
"Vehicle to everything" (V2X) communications, 19
Vulnerability, 2, 5

W
Wannacry "ransomware," 1
Wi-Fi technology, 20
Wireless interfaces
  long-range wireless communications, 18–20
  short-range wireless communications, 20–22
WPA2 vulnerability, 2